IT MANAGER'S GUIDEBOOK

A Starter Guide for Aspiring Leaders in the Technology Industry

Voytek Andrzejewski

For my wife, without whom, my ideas would end up no more than a distant memory.

IT Manager's Guidebook

Contents

Why I Wrote this Book

Hello there. You have probably picked up this Book or listening to it because you are either about to start a new role as an IT manager or keen to see what someone who has done it for many years may offer.

This Book is based on my real-world experiences in this field. I have learned skills from some good managers and understood the mistakes of other not-so-good ones. I have made poor choices on many occasions and tried different strategies, but I always wished there was something like this Book available for me. Something that came without all the lingo that tends to come with management books.

I started from quite humble beginnings. In my school years, I really had no idea what I wanted to do with my life. The Highschool I attended was pushing me and my peers toward careers in medicine, law, or commerce. These were subjects I wasn't really interested in. I love video games and technology but never really had exposure to careers in that field at a young age. As time went on, I realized I wasn't going to amount to much through high school, so I left to start working. I started cooking and became a Chefs apprentice. While I did well there and realized I had a good work ethic, it wasn't a passion for long. Around the time I was finishing up my fourth year as a Chef, the Army was advertising heavily here in Australia, so I signed up. There was a wide range of roles you could choose from, and thanks to a conversation with my mother, a path in Signals seemed to have a future. She insisted I was clever enough to try for a Signalman role, dealing primarily in LAN and WAN systems. I gave it a shot, sat the tests, and passed. This was an amazing experience, and I was finally learning about IT systems. My passion had been unlocked.

I am not saying that you must have a passion for IT, but it certainly helps. When times get tough in your career, leaning back on your passion will be your crutch. This passion will keep you going and moving into better times. It helped me, and 20 years later, I am managing teams of passionate IT engineers and admins using everything I have learned to promote their development. You can do the same, and by reading through the steps in this Book, I'll show you how.

I ended up where I am today purely through hard work and commitment. I stayed true to myself and respectful the whole way up the ladder. I changed jobs a lot, but only to continue challenging myself. This is important. If you stagnate in a role or feel bored, it is time to seek challenges. As time progressed, I found myself taking on roles that managed people and business processes, and I was so lost. I read some books that made no sense, others that didn't fit my situation, and some that were just plain wrong. I learned, through trial and error, from experienced people willing to share and great forums of like-minded peers. My goal here is to target new managers or promotable people that need some starting points to work towards.

I am not here to put you to sleep with graphs and processes to make myself look good. I just want to relay what I have learned and hopefully give you a different perspective on the actual reality of the job.

The IT manager role means a lot of different things to a lot of different people, but I think the basis is the same. IT systems need to run and support the business, no matter what it is. Your job as the manager of these systems is to keep them running and running the best they can. No need to mess about with justifications for what you do. Your job is to keep these systems up and performing. Always remember this.

I want to help you do this by providing some easy steps. This Book is about those steps and an attempt to arm you with some knowledge on what to do in your first day, week, year, and eventually how to maintain a smooth-running department.

The coolest part of our job is we get to see technology do amazing things. I have worked in mining and seen systems that monitor a pile of dust turning into a bar of gold. I worked in finance, where systems monitored second-by-second transactions. I have seen racks and racks of equipment turn into a single rack yet provide the same experience. I love this work, and I hope you do too. You should keep that passion because now you get to lead teams that will notice the passion you have. You are not a manager for the sake of a bigger pay packet, you are stepping into a leadership role, and as such, you are going to drive a team to do great things.

Let's delve into what to expect and what you should focus on to be successful. Sure, add your own flair and read other books, but I want you to have a read of this Book and come away realizing that it's not overwhelming, and you don't have to be Superman to do it. This job should be fun, and I'll help you keep it that way.

Your Roadmap

Let us look at the book from a very high level. In this section, I want you to understand this roadmap is a guide for you and to highlight the points you should focus on. Additionally, I will break down when you should focus on them. The book will delve into these a little more as we move on, but for now, just take a moment to understand you do not need to understand everything right away. You can plan your actions over a longer period of time.

When you start a new job or get promoted into a management role, it is very easy to get overwhelmed by the status quo. You know, the whole "this is how we do it" mentality. You quickly find yourself falling into line and lose sight of what the problems were in the first place. After a few months, you are a piece of furniture, and all the aspirations to change are lost to complacency. While this is a comfortable place to be, it is not the intention of your role. People hire managers to introduce positive change. To reflect and improve. To evolve and enhance. Use this book to help you do that. There are many ways to function as a team, but the steps here are to help you reflect on how your team should function, improve their performance, and evolve the team into a well-oiled machine.

I wrote down these steps personally without all the wording to keep myself in check as I grow with the company. They are my personal compass to ensure I don't end up complacent, and I worked hard to achieve a lot of the guidelines detailed here. I want you to do the same. In fact, I made a similar checklist at the end of each section for you to use. Tick them off as you go and get a sense of accomplishment when you have implemented them.

I follow this guide every time I take on a new role and focus on these areas before moving on. Sure, they may chop and change depending on the requirements of the job but not dramatically. Keep referring to them and take stock of what is missing from achieving them. Understand the gaps and go at it from a different angle.

Book Overview

This book is broken up into sections based on a loose timeline. The key is to achieve the goals set out within each period before you move on to the next area. For example, I will lay out the requirements you should look to achieve in your first week at a new job. Try to avoid deviating from these too much or overreaching. Use the time effectively and get to understand your goal. I advise sticking to the goals within the timelines and nailing them; you will soon find that some are harder to achieve than others.

First Week – Meet your team and understand the business.

The first week is critical to your future at a company or as a team leader. You will be judged, questioned, and challenged from the outset, and knowing what to do will help you stand up to these challenges.

First Month – Understand your budget, operational requirements, Basic cyber security, communications plans, and contracts.

Dig deeper and get to know your department's financial goals and processes.

First six months – Set up ITIL practices, key performance indicators, asset management, training plans, and infrastructure resilience.

Start making positive changes to your department. Let's get that machine oiled up and running smoothly.

First Year – Implement standardization, roadmaps, compliance, some more ITIL and SLAs

Nail down the processes within your department and start measuring performance and forecasting expenditure.

Year on Year – Begin auditing and push for growth.

Help your company grow through the technologies you manage.

Preparation

Is Stepping Up the Right Thing for You?

It would be best if you asked yourself this before taking the leap. You are probably in a position where your managers pressure you. Maybe someone in the position has left, and people are looking to you. Or you are in line for succession, and you feel you are ready.

Each of these situations is similar. All of these situations are an opportunity, a chance to challenge yourself. As I said earlier, you should look to new challenges if you are stagnating or bored. That doesn't mean stepping into a management role is a good idea. You need to understand that it is a different set of skills required. Your technical skillset will help, of course, but it needs to take a back seat. Your preconceived ideas of being a good IT worker also need to be left behind. You must understand you're not there to turn everyone into a version of yourself. Empathy is a huge part of your new role. Putting yourself in other people's shoes, seeing the bigger picture, and understanding the business goals, are the focus of your new role.

This may mean you will need to take a back seat and let others do the tech work. Are you going to be able to let go?

This may mean that your technical skills take a back seat to administrative work. Are you willing to work outside your comfort zone again?

This may mean that you need to constantly look for improvements in business processes that don't necessarily include technology. Are you keen to understand financial and human resource challenges?

If you ask yourself these questions and the answer is yes. Go for it.

Let me explain based on experience. When I started, I was very much a technical resource and was expected to configure the usual IT systems one would expect. I reported to management on my tasks, and life was simple. I had a very narrow field of responsibility.

I'll be honest, back then, I felt like my managers were useless and a waste of time. I had no empathy for them, and I didn't get it. I felt like they didn't really understand my narrow little world to the same degree I did, and therefore, their position in the company didn't matter.

As time passed, I was asked to step up and look after a small team that did a similar job to mine. I took the job without hesitation and without thinking it through. It was a hasty decision, and I took a significant risk without knowing it. Luckily, it was a mild step up, and I had worked with the team previously on multiple projects. The role was different from what I expected. My manager put me in the position, and I was suddenly accountable for a budget and dealing with resourcing issues. It was a small couple of tasks, but they made me realize that once you step into this world, it doesn't matter how technical you are; if you want to be good at it, you need to leave that behind. I suddenly had two jobs to do...

Fast forward to today, and I don't even touch the technical aspects of the job anymore. I am interested and love listening to my team working through a problem, but I am in the back seat regarding that stuff. My job keeps me so busy managing the operations that I am losing my skills at a technical level. I don't have the time to scratch myself little own keep up with the industry. It's funny; I feel like the junior guys in the team feel exactly about me the way I did about managers back in the day.

I am sure it's not too late to get back on the tools. But I wouldn't even dream of it. I love my job, and my new passion is making sure my team has what they need to do their job, ensuring the process is more efficient, and looking at the bigger picture.

So, going back to whether this is right for you, think of these statements. Management in the IT industry is a different type of work, so take a good look at yourself and work out if it's the right thing for you. If it is, keep reading through it, and I'll give you the tips to get it done.

If you want to make the leap, let's start at the beginning...

Change Your CV

Let's start with your CV.

Right now, it's probably highly technical. You are a CCNA, an MCSE, a Cloud architect, etc.

All great stuff to put on there, but this is no longer front and center. Your ability to lead needs to be. You need to show potential employers that you are comfortable leading people, being a "go-to" for escalations, understanding of the business process, and you can manage a budget.

These values need to be front and center. Make sure you place these items within your previous role definition. Include numbers, percentages, metrics, etc., that cover your knowledge of these fields.

If you have mentored people, led a small team, had to balance the books, and were an escalation point in your company, put these on there as well. Detail how you improved a business process. It may be as simple as creating a workflow to set up users or improving a ticketing system by generating reports for your boss. These are all great bits of info people will value.

Here are some examples.

- Mentored and managed scheduling for a small team of 3 IT professionals across two sites

- Implemented an annual auditing process for company IT assets covering over 1000 devices for improved lifecycle and warranty management.
- Designed and managed a $100K upgrade project to lifecycle legacy systems.
- Brokered a mobile contract that lowered the telecommunications budget by 10%.

You need to think back through your career and find good examples like these of how you have improved the company through more than just fixing printers and creating user accounts. IT management is about benefiting your organization at another level, and if you are aiming to take this leap, you should have some exposure to work like this. So, dig deep and rack your brain. Get these examples on your CV. Polish it up nicely and keep your technical skill set in the skills matrix at the back of the CV. Front and center need to be these great examples of leadership.

If you can detail specific metrics, this goes a long way to show your understanding of a business mindset. Detail a percentage here and there. Explain cost reductions through actual dollar figures or show growth by detailing headcount.

Don't forget to update your LinkedIn profile as well. While your CV will be handed around, many recruiters will check your LinkedIn profile and look for consistency. So be consistent. Change your profile to match up with your new CV changes. Update your position from Engineer to Manager. Update your picture for a suit and tie. All the details need to be squared away. You don't want people to have something to pick on. You must be perfect and meticulous in your CV and profile.

Now it's time to take your CV and get it out there. Talk to recruiters and seek their advice, jump on job sites, and apply for roles you may be ready for. Don't be too scared of some of the stuff companies ask for. Sometimes they just add fluff there that may not be relevant. I can't count how many times I have sat through an interview and realized the job posting and description were nothing like the job!

Most importantly, give this process some time. Let your CV flow through the different channels within companies, and before you know it, you may find yourself being called to meet a hiring manager. It can be extremely discouraging to be knocked back time and time again. This happens to everyone, and it's not one thing or the other. It's nothing personal, so don't take it to heart if this process takes a while and there are many rejections. Remember, success is achieved on the other side of failure.

The Interview

So, you have landed an interview. You may be going for that IT management position because you feel you can do a better job or think it is time to challenge yourself in other ways. This is good, and I know it sounds daunting and a bit scary, but trust me. It is a fulfilling next step in your career. Go from being the in the engine room to being in the pilot seat.

The fact that someone is calling about the position means that you have done all the right things on your CV, cover letter, and LinkedIn profile. I gave some tips on how to do the CV properly, but there are experts out there that can help you more with that. Doing your research is an excellent idea because an IT manager's CV differs significantly from a technical role CV. And a good CV is a tricky art to master.

Anyway, you have nailed an interview, so let us dive into that.

Be sure to research the company looking to hire you before the interview. See what drives their business, where they are located, and if you can find the technologies they use. Anything and everything about them from a Google and LinkedIn search. Be prepared to know a little about who they are and what they do. I always ask candidates what they know about the company they have applied for, and let me tell you straight, it doesn't sound good when they have absolutely no idea. You should have some background knowledge of what your potential employer does. The more info, the better, and the best scenario is you have questions prepared for your interview on anything you can't find from internet searches.

Your first contact may be a recruiter who understands the job's general terms and not much else. Let me tell you one thing. Not all recruiters are not out to help you. In my experience, I have only ever had one or two recruiters that cared. The remainder were out to fill positions as quickly as possible and ensure their client (not you) were happy with their choices.

The fact that a recruiter has called you is good, but generally, they want to hear who you are and your salary expectations and that you would be comfortable with the high-level particulars.

You want to answer the recruiter's questions and stick to the fact that you can help their client. Always sound eager, even if it is a company that may not be ideal for you. Interviewing is about getting time up in the hot seat. The more you sit in front of recruiters or potential employers, the more you learn and the better you get at pressure questions. So, take every opportunity you can to be interviewed. When you finally get the interview for the perfect job, you want to be ready, cool-headed, and able to respond articulately. I was blindsided many times in interviews and found some excellent questions I was unprepared for. It took me many interviews to finally be confident enough for anything thrown at me.

After the recruiter is happy, you may get hit with an HR department asking similar questions. If HR calls you, it usually means you meet their criteria, and they want to tee up a good time for a chat with the hiring manager. Also a good sign, but you will need to prepare for the most important part. The hiring manager.

The hiring manager may be the company C-Suite (CFO, CTO, CIO, etc.) that interviews you or the head accountant, director, or ops manager looking to fill the role.

These guys generally have a preconceived idea of what an IT manager should be. Your interview should be spent working out what that expectation is. You should ask questions that probe that expectation.

For example.

- What would you expect me to achieve in the first year?
- Do you have any issues with the running of the IT department?
- Am I replacing anyone? Why did they leave?
- Who am I reporting to?
- What would mark my position as successful?
- What are the challenges you hope I can overcome?

Basically, you should angle your conversation on how you can help.

Obviously, you will need to explain how you fit into the company and what your management style is. If you have done your research, you should be able to answer these truthfully and give them what they want to hear.

Always use examples of how you solved problems. Solutions to problems is what your job is. As an IT manager, you should advocate for positive change and provide direction. If you cannot go back through your career and find the solutions you have come up with and, more importantly, lead, I am not sure why exactly you are in any position to apply as an IT manager. You need to be able to go back in your history and find the stuff you did that provides a positive direction for the companies you worked at.

Arm yourself with these memories and take them with you to the interview. Fire them at your interviewer. Dazzle them with your positive energy. Detail your incredible foresight and that the money they have on offer is probably not enough for just how amazing you are.

If you see the interview going well and have some time to go over further details, see if you can find out the below points.

- What systems do they use?
- What systems does the company rely heavily on?
- Are they in the cloud or on-prem?

- What does the company sell or do?
- How big is the company? How many users can you expect to support?
- How big is your team?
- Do you need to take over any projects that a mid-flight?

A key point is to never, under any circumstances, bad mouth your previous managers. This will put off your potential employers, especially as the first thought in their mind is you will say that about them. Just do not say anything negative. If you disagree with the previous direction, state how you dealt with it in a positive light, not abuse the last person that provided the direction.

Success Can Take Time

Hopefully, within six months of working at it, someone is willing to offer you a job as an IT manager in their company, and you are popping the champagne with friends to celebrate. If not. Keep at it. An important fact to note, which can easily be forgotten, is that each organization you deal with is different and will see you differently. Let me explain.

It is very easy to become depressed and think that no one out there wants you because of you. This is far from true. Failing multiple interviews doesn't mean they all see the same individual problem, and you go crazy trying to figure it out. You need to think that each organization has individual reasons for not choosing you. For instance, the interview with company A was perfect, but they chose a more experienced candidate. The interview with company B went badly because you didn't prepare. The interview with company C went well, but they hired internally. All different reasons.

You can't sit there beating yourself up over these differences. Follow the steps I outlined above and just keep at it. I always look at the jobs I didn't get as a good sign, a sign that I wouldn't have been happy there anyway. Trust me when I say it's harder when you are successful in multiple roles at once and need to make a choice. So, my advice is to chin up, move on, don't beat yourself up for six months of trying. It will happen, especially if you have that goal to step up.

Anyway, however long it takes, you have taken the first step, and once you are successful, pick this book up again, and let's dive into your actions when you get the call to say you have the job!

Preparation is Key

It's been a bit of a journey to get to this point. Getting your foot in the door took a lot of hard work, but it should have paid off. You are hired, and all that effort wasn't wasted. So now that you have the job in the bag, it's time to get squared away on what you must prepare before your first day.

First things first. Hand in a resignation letter. There is a lot of info out there with templates etc., but I advise keeping your letter brief, keeping the essential dates on there, like your last day and today's date, and handing it to your current boss the same day it is dated. Give it to him or her in person and let them know an amazing opportunity awaits. You don't have to disclose too much if you don't want to, especially as it may be an emotional time.

Once the resignation is in, you have a few weeks of your notice period to plan and prepare. Don't waste it. Spend some of your spare time, and I mean spare, don't burn the bridge with your existing employer to get a few things squared away.

Firstly, go out and get suitable attire for your new job. I follow the rule of "dress for the job you want, not for the job you have." It's a simple rule. Right now, you may have stuck to polo shirts or maybe a business shirt with jeans. It doesn't matter.

You are about to start a leadership role, and this is the first step. Leaders dress well. Business clothing is appropriate, a suit even better. Remember that you will be the face of seniority in some aspects of your new company. Dress for it and get used to looking good. It will be your second skin in the future and a great first impression for anyone you meet. You don't want to visit clients or your new team dressed like one of the engineers.

You may be thinking suiting up for this role is over the top. The author doesn't understand my new job etc. Rubbish… It really doesn't matter what type of management role you take. Simply by looking professional, you have already improved it. So go out and grab yourself the appropriate gear for the job and look smart.

Next is to do a little spy work or investigation. See if you can find out everything relevant to your position. You may have covered these points during the interview; if not, maybe you can email them to your new manager or see if they can take a call and answer these. Impress them by stating this is all part of your preparation for the new role.

- What systems do they use? Anything outside of your experience?
- What does the company rely heavily on and require max uptime?
- Are they in the cloud or on-prem?
- What does the company sell or do?

- How big is the company? How many users can you expect to support?
- How big is your team?

As I said a few times, you may have found most of this information in your interview or before your interview. Just make sure you understand the company's general goals, especially what they are trying to achieve. Your job is to help them achieve it. We know that IT is the backbone of any company. You will need to make that a reality and not a hindrance.

Also, make sure any technology the company uses is something you are comfortable with, and if not, get busy learning it.

You get a job at a finance firm that uses SAP. Well, get busy learning SAP. Get a job at a company using Office365; guess what? Watch all the videos on office365 (in particular, the licensing process!!) You get the picture. Try to walk into your new job knowing the tech they use so you are not blindsided about them in your first days.

The First Day

Onboarding

Well done, you are getting through the first steps nicely. You have landed the new job, researched the company you will be supporting, and hopefully studied up on the technologies you were previously unfamiliar with. Maybe, you even brushed up on the stuff you already knew.

Time to head in for your first day. It is a big day and incredibly important. I call this the "first impression" day. You need to be 100%. Not trying to put too much pressure on you here, but the first time people meet you, they will judge you immediately. As a manager of people and, in some respects, a company leader, people will look for anything to nitpick on.

We already talked about the appropriate attire, so dress well. Look the part and stay professional. You will meet lots of people, and they will see you for the first time. Smile, even if it is unnatural to you. Look people in the eye, be honest in your answers, and be true to your questions.

When you meet new people, do whatever you can to try and remember their names. This should be your focus on the first day. It is always super impressive when someone remembers you, even days or weeks later. It's something I noticed about effective leaders. They always remember your name and something about you long after your interactions.

So, a little tip if you are like me. My memory tends to fail me at the worst of times, so I do a little trick to remember people's names. When you are speaking with a new colleague, try to use their name in a sentence immediately after they introduce themselves. Make a conscious effort to reuse their name as often as possible without sounding weird. It does not sound like much, but this triggers something in your brain to store away the name. A simple thing like remembering a name goes a long way to gaining people's respect.

Having said all of that, you probably won't always remember every name. On your first day, do your best to remember who's who and what they are responsible for. You may find some people that seem to be responsible for nothing. Make a note of them too. I have found that these people, when they are bored, they are usually the ones that ask for everything and complain when it is not to their expectations.

The first day is generally spent seeing your main office area and facilities. The onboarding process can take a few hours or days if the company is large enough. Just enjoy it while it lasts and try to introduce yourself to everyone. Do not make any judgments about anything. Take it all in and enjoy the fact that you are meeting new people. If you find anyone with common interests, note it, as they can become your friend in the coming months.

Remember that it is super important not to make any judgments or comments about the current state of the business, processes, or systems. You may be commenting on something that was decided by those around you. You will quickly make enemies if you start making statements about the state of things. Just hold your tongue, and enjoy the scenery as your time will come to make a positive change.

You may be asked to join meetings with other leaders of the company. In these instances, I find it is best to just listen to what is said. Ensure you make no judgments. If asked, "What are your thoughts on X or Y?" remain neutral as you may be trapped into picking a side. Unless it's technology-specific and you know that X is better than Y, keep yourself out of the internal politics for now.

Make sure you ask plenty of questions. In my opinion, there is no such thing as a stupid question. In your first week, seek understanding and soak up as much information as possible. Work hard to ensure your questions are not condescending or judgemental. Sometimes even simple questions can upset people. Focus on getting information about processes and details, not people or decisions.

Try asking questions that show impartiality, like "What is the system used to onboard customers?" and avoid questions that can patronize, like "Why did you choose that system?". There is a distinct difference here, and it will be important on your first day to ensure you are making friends, not getting people's backs up.

The main aspect of your first day is to soak it up and meet and greet. You want to do this confidently and gain as much respect as possible from those around you. Always remember where you came from to bolster your confidence. You may have spent many years in the trenches to get to this point. Use that to remember you deserve this position, and they made the right choice. Self-doubt may creep in on your first day but don't let it. You just need to meet and greet. Focus on that.

Side Note - The Culture

Here is an important piece of information. On your first day, you will no doubt meet the team and those you will work with either closely or partially. It may be one of the few chances you get to talk with people outside your area of influence because once you get started, you may be focused internally for quite some time. So, as you are meeting people and wandering around, try to note the culture in the office.

- Is it a quiet place of work without any camaraderie?
- Maybe it's a bit of a boy's club?
- Maybe it's the OK Corral filled with Cowboys?
- Is it a utopia, or is it a toxic cesspool of negativity?

You need to gauge how the work is being conducted and, more importantly, if the team works well together and how the organization treats them.

I would almost always prefer a workplace with overt communication at its core to a place of quiet solitude.

To me, the fact that people are talking amongst each other to deal with the day or spend part of their time shooting the breeze tells me that the culture is good. Silence among colleagues is a sign of dysfunction.

We are not robots, and I truly believe enjoying being at work is critical. Nothing wrong with having a good time while you are maintaining professionalism. A healthy culture is one where people enjoy themselves and coming in on time because they like being there. Sure, it is a job, but remember, these people spend as much time here as they do with family. Sometimes more so. Very important that they are happy to come to work and do their best because they are happy to do it. Not because they are forced or feel they must.

Of course, it can be the other way. You find that your team seems to have too much time on their hands and spend most of the day sitting around chatting. This is also not a great sign. While it's great that people are enjoying their time at work, there is a business to run at the end of the day. Everyone is here to earn a paycheck. The key word is earn. No one is entitled to free money in this world except the royal family, so having a group of people spending most of their time doing nothing is an opportunity for you. Make a note of it on your first day. Whether it's a productive work environment or maybe a playground, you just note it here so you can change it as time passes.

The perfect workplace to me is one where the troops are hard at it, communicating well, and the odd smile or two crosses their faces. You should aim for this.

The First Week

The First Week Plan

Now that you have a feel for the new company and know the people, it is time to dig a little deeper. Again, although it may be hard, we will spend the next few days observing and noting how everything works. This is where your value to the business will start to take shape.

Use the steps below to guide what you should focus on in your first week for your new role. These are only guides, and I want you to cover these steps. Don't sweat the timing, though. It may take a couple of weeks or a month. What is important here is that you cover these points before moving on in the book.

You will be required to conduct business as usual (BAU) activities and maybe write some reports, so spend the time to learn the current process and own it. It may be a process you do not agree with or want to change but follow these existing processes to the letter and nail them. You must remember that the business was operating before you stepped through the door, and while it can do better, it still needs to function, and your changes should come slowly.

While learning the day-to-day processes for your position, you will also begin a plan to best position yourself as a company leader.

Meet the Team

Your first day or two was looking around the office, touring the facilities, checking out the local lunch bar or café, and meeting with the right people. It is time to get to know your team a little better.

These guys and gals are your engine room, and, as the captain, treating them the right way will pay dividends. They have the power to make you look good or like a fool. It is incredibly important that they get treated with the respect they deserve. If they deserve it that is, we will get into this later.

A good start is having each of your staff take you through their area of expertise. I find it a good trick to play dumb during this phase. It can be hard, and you may want to show people up with your big brain and knowledge, but just play is dumb for now.

Ask lots of questions and try to unlock their passion. For instance, you can ask your network specialist to take you through their setup. No need to dumb down too low by asking, "What's a router?" but try to see what makes the network admin tick by asking about the type of security appliance being used to face the internet or what access switch is being used to feed the workstations. You are trying to get them to open up about the gear at a level that excites them to talk. I find playing dumb is a good start. "I see you are using all Cisco gear; I have had problems with it previously. Do you have any?"

Hopefully, they will disclose their problems with their equipment, their preferences, and, more importantly, their plans to fix known issues within their area. Take note of these plans as they are a goldmine for a good IT manager. Note, collate, and devise a strategy to deal with them. We will go into these later in the book, but for now, make sure the problems faced by your team are at the very least noted and you are showing that you empathize.

Make sure you spend time with each team member and get to know how they feel about their place of work. Find out if they feel anything is missing or if there is anything you can do for them.

Most of the time, these initial discussions should be quite positive and informative. However, you may also find yourself listening to constant whinging about the company, the department, or about other members of the team. This is a sign of a toxic culture, and you, my friend, have some work to do. Dealing with toxic cultures is incredibly difficult and many managers do not handle the situation well. Ignoring it does not help, but I will guide you through the steps needed to fix a bad culture, at least at your department level.

Remember what I said about the culture in the previous chapters. Well, the IT department has varied characters, but one thing that brings everyone together is our love for technology. This common ground makes it easy for IT managers to glue their teams.

While you may find yourself in an environment that is quite toxic or has negativity at its core, do not stress. It is important to note this is the case, but following the steps further down, I have successfully converted a poor culture into a vibrant, professional, communicative one within six months, and you can too.

Note the Systems

As you are shown around by your team members or discussing their various fields of expertise, you need to start making notes of the systems they support. You need to know what makes the organization tick and what technologies are at the heart of it. A key point to understand is if the team is looking after a wide range of technology or sticking to a select few. Have they standardized on certain systems or picking and choosing at the whim of certain individuals? Are there any systems outsourced?

Either way, make a note of all of it. You need to know all the technologies you and your team will support.

Create a Skillset Matrix

Another big part of the process is determining who can do what. And how well they can do it.

Ask HR or your staff to provide CVs. These may not cover the full skill set of the member they represent. Remember that not everyone knows how to explain their capabilities very well. So, use the CVs to understand the fundamental skills as a starting point. You also should understand the technologies keeping the business running. Remember that you did this during your meet and greet with the team members. This information is incredibly important for creating a skillset matrix to keep the lights on.

Focus on the technologies running in the business and need to keep running. Marry up these technologies with relevant skill sets. You can use my example below, which I use to clarify the team's capability.

Start with the technology or system and move across to the skills required. You will need to expand on this matrix as time progresses, and you will have to, but for now, you must thoroughly identify the skills required to keep the place running.

Name	**Systems**		**Networks**		**Languages**		**LOB**
	Windows Server	*Linux*	*Cisco R+S*	*Cisco Wireless*	*Python*	*Visual Basic*	*Arrow*
Jane	Certified	Certified	Basic	None	Expert	Intermediate	Expert
John	Intermediate	Expert	Certified	Certified	Certified	Expert	None

The table shows that the technologies are broken down into four categories. Under each category, you need to identify the technologies running in the business. As a small example, I have added a couple of basic ones under each category. Start with something like this, relevant to your workplace. Fill out the table with everything running in the business but do not get too granular. Windows should be enough to cover things like DHCP and DNS, maybe even Active Directory. But not Exchange or SQL. That would be separate skills like databases that should be separated out for your DBAs (if you are lucky enough to have them).

Names go down the left side, and then you simply fill out the data.

As the CVs will not give the level of detail you need to get this matrix 100%, it is important to have your team self-assess. Allowing your team to note their skills will help you understand some things. Firstly, the main reason for putting this together is to understand if there are gaps in the technology that runs the business. Secondly, the level of confidence your team has in themselves when they assess. You may find that your team members already begin to exaggerate or downright lie to you about this. You won't know yet as you just met them, but you are at least keeping a record.

Give your team the following key to assess themselves.

Knowledge Levels	Certified
	Expert
	Intermediate
	Basic
	None

It's a straightforward chart that covers the levels of competency. You may think being certified is exclusive to being an expert, and I agree. However, the industry and partnerships with our suppliers benefit more from certified people than highly experienced people. A bit of advice from someone who has dug into this phenomenon. Certification in IT does not make an expert by any means, I have met people with certifications across many fields, and they still have no idea how to deal with real-world problems. BUT, and that's in big capitals, certifications can unlock and qualify your IT department for discounts, gold partnerships, opportunities, and winning bids. It's just the measure of a company's IT proficiency, and all the experience in the world may not matter if you have zero certified staff.

Once your team has completed the matrix, you should review and questions any that do not make sense.

For instance, a network engineer in your team may have entered intermediate under networking, and they are your only network engineer. You need an expert right? So, speak to them. Ask them why they feel they are not an expert, and you may find they just undervalue themselves or were placed in that position without their consent. They may have even taken on that job because the company needed them to. They need your help, or they need to be replaced. This is the point of doing the matrix and questioning the skillsets. You don't know these people at all, and this helps you find those gaps.

Once completed and you have done your review, you should understand the technologies covered by your team. When you collate all the information, the final step is determining which technologies have a gap. For instance, if you notice that the company relies on wireless technologies, and you have one staff member scoring themselves as an expert and everyone else basic or worse, you must target that technology ASAP.

It may sound like a horrible way of thinking, but you need to start remembering that any member of your team could win the lottery or leave the company. You do not want to scramble to find a like-for-like replacement after this occurs. You want to be prepared for this eventuality. You do this by identifying the gaps and closing them.

We will deal with the gaps in later chapters with development plans for your team. I don't expect you to fix this problem in your first week, but you are already identifying ways to improve the team. Good work.

Find the Gaps

By now, you are starting to understand what each team member looks after and is passionate about. This is a great start, but you need to have the missing pieces in the back of your mind.

There will almost always be a piece of technology vital to the company sitting idle and unsupported by the team. Whether it was implemented by shadow IT (another department within the company that runs isolated systems) or an ex-employee, there is a good chance there is something that is missing the care and support it needs.

A SharePoint server no one wants to know about, or a POS system for the back-room bar at the hotel are just a couple of examples of what I have seen. You must take these seriously when you ask the IT team, and they do not want to know about it, but other staff would be lost without it. Findings these sorts of gaps are incredibly important and will assist you with being a good manager of IT.

You need to identify these loose systems as soon as possible. As the team has shown you the technology they support, so should the end users show you what they rely on.

When you are discussing, company systems and a system are mentioned repeatedly. One that you cannot see anyone actively maintaining or supporting. Or you are told a system was implemented a few years ago, and no one knows how to deal with it. These are red flags for you. Your team may not think it is their job, but you will make it their job. The company may believe this system is fine, but you know having an unsupported system operating in-house is not fine.

Always remember, it is your job to be an empathetic ear between your team and your users/customers and ensure that your team is meeting the users' needs. Remember, without the users; you do not have a job. Always try to get their needs met to the best of your ability.

How do you find these unsupported systems? Well, it will only come about by asking the right questions. Hopefully, one of your team will point it out during your walk around, or a staff member will mention some system they use that was not mentioned by your team. Either way, you need to find these systems and make a note of them.

Set up Spot Meetings

Get your team to start committing to regular reporting. Some managers are standoff guys sitting in their crystal towers, shooting out updates via emails or instant messages. That is not good leadership, in my opinion. You should be in front of your team regularly; meeting them at the start of each week is the best way.

These are a few different terms, but I like to call them spot meetings.

These are quick, one-hour sessions with the group. Split them up into several groups if you have several teams. For instance, I have had teams across several states, each with unique issues, so I split the meetings into 1 hour each among those teams.

Once you have split up the teams, have them meet with you each week. Set up a calendar invite and make your team aware of the importance of these. Set the meeting to the start of each week and make attendance compulsory. If a member cannot attend, they must let you know beforehand, and it must be a valid reason, like leave or working for a customer. Make it a KPI if you must but ensure that your team understands these meetings are vital and their attendance mandatory.

Once you have established the schedule and clarified the importance of the weekly meeting, set the agenda to be fixed for each member to explain their weekly tasks. You can lead by example and begin the meeting with any updates you have. What new information is relevant to the team and the technologies being used, safety awareness, HR updates, budget constraints, and what you are working on. The main aim here is that everyone on the call or in the room explains their plan for the week.

Your team must provide tangible goals that can be achieved for that week. If people start blabbering about some project that may start next month, stop them and bring them back to the week ahead. Ask them their goal this week and how they will achieve it. Simple.

Once you are done providing information, go around the room and get each member to pass on their update to the group. Before you know it, the guys will start opening up about issues they have seen or successes they achieved, and this communication is pure gold.

Do this every Monday or whenever your week starts, and have your calendar invite recurring.

Your checklist for the first week

So that should be roughly the first week done and dusted. It was a busy one, for sure. You should have been a sponge and asked a lot of questions. There should be no judgments made or comments about how the place runs. You are gathering data in that week, which may extend to more than one. This book is a guide, not a stringent set of rules, so be flexible in how long it takes to gather this information. Remember, there will be aspects to the job that your new employer will need you to pick up immediately, so spend some time ensuring you nail that as well.

Let us go over what you should have completed, and if not, allocate some more time but get these squared away before moving on with the book.

- Meet your team.
- Spend some team with each of them individually.
- Make notes of your team's skills, strengths, and attitudes.
- Start identifying the gaps.
- Set up regular meetings.

The First Month

Now that you know your team and their capabilities and they are getting to know their manager is interested in them, you are off to a good start. It's time to understand the key factors that set you aside from your team. Remember, where they run the engine room, you drive the ship.

From an IT perspective, that has nothing to do with pushing the right buttons anymore and everything to do with making sure those that do push the buttons have what they need to push those buttons.

Side Note - The Imposter Syndrome

I think it's important around this time to discuss a phenomenon known as the imposter syndrome. You may feel it in your first month or even during your first week. It's a terrible feeling I go through every single time I take a new job. Self-doubt creeps in at the edges of your mind and makes you question yourself and whether you are worthy or qualified for the job. It's not a great feeling and can lead to problems. When this self-doubt starts to creep in, it can lead to depression, poor decisions, and hurt your credibility. I want you to know that imposter syndrome is perfectly normal. It is a problem most people face, a sign that you are a good manager and will be one of the best.

My tip when you start to feel this creeping in and question what you are doing is this. Acknowledge it for what it is. A perfectly normal reaction to leading people and challenging situations. Once you acknowledge that this is just a reaction, you begin to understand better that it is something you can overcome. The next step is to get over it. I do this by remembering where I came from.

You have a wealth of experience in this field and much to offer. Once you get through this book, you will have many goals to achieve, and they will all strengthen the business. So, stick to these facts.

To reiterate –

1. You are experienced and worthy of your role

2. You have goals to achieve, and they are valuable.

Remember these facts, and the imposter syndrome and self-doubt will leave you alone to get on with it.

Determine Your Budget

Let's kick off your first month by understanding your budget. You must work out how much money you have to maintain your world. An IT manager's success relies (partly) on the ability to know what the company spends on IT. You should understand this aspect quickly, how the money is spent, what is the process of spending it and how much control you have over it.

I like to split up the budget into two major portions.

OpEx – keeping the lights on.

CapEx – Modernising and improving.

Let us look at these in more detail, but first, you will need to find out how much of a budget you look after.

The best way to do this is to go back to whomever you report to and ask the question directly. What is my budget? Do not let up on getting a solid answer. A good manager should know about this portion of the business. Your other job is to save your organization money wherever you can or get more value from your spending.

I have had many answers to this question, but they are usually given in the form of a spreadsheet with a predetermined amount, or the answer is "Don't worry about it."

Suppose it is the latter, great. It makes life easier for you, but it means you need to beg for money when you want to achieve anything.

If the answer is a spreadsheet, you need to get across that quickly and thoroughly. There may be a lot of stuff there that makes no sense or has strange titles. Remember that this could have been a way for your predecessor to hide spending. Line items like "medium transmission costs" or "Tertiary commissioning expenses" should be called out and questioned, and you will need to fix that quick smart.

Take the spreadsheet and break it down into items that make sense. Note the monthly costs and a yearly total at the end of your budget. If you need to break these down into subcategories, so be it but ensure you have these main line items in bold with the totals visible. This will help you when discussing next year's budget with senior management.

I usually break mine down as below.

OPEX

OpEx is your Operational Expenditure. The particulars that need to be paid to keep the lights on. These will include major items like;

WAN services – This refers to any internet, WAN, MPLS, or Wireless links required for your sites to talk to each other or talk to the internet. Generally, these services are rendered under a contract period and have a monthly cost.

Mobile services are the running expenses of staff mobile phones or other mobile internet systems.

Hosting – This is all your "as-a-Service" stuff. Anything cloud-based. They should be noted here if you are paying for hosted servers or 365 services.

Licenses – This is all the licenses required for your on-prem gear. Hosted cloud licenses should be included in the hosting section, but here is for all your on-premises stuff.

Hardware maintenance – Desktops, printers, servers, and the like may have a cost associated with them, and leasing plans may be in place for your equipment. These need to be noted in this section.

That is it. You should be able to fit all your operating expenses into these categories. Do not create a column for "Other." If you find that you are paying for anything outside these categories, you are probably spending on items outside your budget, and you should have them adjusted out of your budget. Remember that the OpEx budget is for keeping the lights on and your systems running with good warranty backing.

Once these are categorized correctly and checked over historical spending, you may notice items that can be migrated into someone else's budget or other potential cost-saving scenarios. Highlight the expenses that look over the top or unnecessary. These will help you kick goals with your boss.

We will get into cost savings later in this section. It's an important part of your job. You will make your department a lean, mean tech machine. Next, let us deal with what is left over in your budget. Once you are satisfied, you have allocated the money for maintaining BAU.

CAPEX

Capital Expenditure. The particulars that move IT in a new or improved direction. I call this future-proofing money and always think that it is for improving your IT resources and systems for the future.

Training – Always allocate a portion of the money to training. Your staff must get away at least once a year and upskill in something. The company may be paying them well, but unlike other fields, the changes in this industry mean your staff needs to constantly learn. So, look at the headcount of your team and look at your budget for the training. If you notice a huge shortage of funding in this area, look to provide time off and a login to a training portal like Pluralsight. Training should not be negotiable. You need to provide something, especially if you have big ideas for improvements, and those improvements require technology outside of the skill set matrix mentioned earlier.

Hardware purchases – Later in the book, you will deal with assets and probably notice some gear is aged or end of life (EOL). Be aware of what amount is allocated to refreshing hardware and prepare accordingly. You may need to prepare yourself to create a business case for additional funding in this area. I will take you through business cases soon but identify where your annual budget aligns with your refresh goals.

Research and Development – Your team and you may want to look at new equipment or technologies every year. Try to squirrel away some money for this research.

Major Projects – So this is the big one. Assigning money to the major projects planned for the year ahead. You won't have created your roadmap yet, but this will likely go hand in hand with what you do in that space.

Think of major projects like large hardware refreshes or new software releases.

You may have joined midway through the financial year, and these projects may already be underway but keep in mind what has been spent and what is left over. If there is a Project Management Office (PMO), chances are, they are dealing with the budgeting and may come to you for more, but you need to ensure you understand the volumes regarding your project budgets.

If you are not conducting major projects or have money left over, consider placing them into your R&D spending and let your team know there is money for R&D., You will quickly get bombarded with requests to try new tech. The next section will help you work out how to ensure you can forecast your spending and control the amount of spending.

Salaries

You may have noticed that I left salaries out of this section altogether. The reason is salaries are generally not shared with you. Usually, the HR department dictates and approves salaries for the staff of the company you work at. They will ask for your input and advice, but generally, the salaries are outside your remit.

Either way, you should keep tabs on what your staff gets paid. HR may tell you it is classified information, but if you want to keep your staff, you should understand whether your team is well-paid, underpaid, or on par with the industry average. All of these are important when it comes time to review salaries or, worse, review head count.

Spending Your Budget

Now the big trick here is when your budget is up for renewal, you will want to have spent the CapEx portion. The secret in the industry is that spending or slightly overspending your CapEx budget is the way to go. You can look like a hero by saving some money here or there, but the goal is always to make sure you have spent what you were given. The sad news is you will lose it if you do not. Bean counters will immediately adjust the forecast to suit underspend, and you will have to raise business cases to gain back the money you lost. So yeah, spend the money if you have it.

Now back to the R&D section. I mentioned you have extra money, and your team is eager to trial new tech. This should be encouraged, but remember you need to make a fair process for the team to do so without anyone feeling like they are missing out. So, what I do is mention, just before the financial year end, that the budget is being revised and that this is the team's chance to bring up new tech they want to trial or buy. Give the team a month to develop their ideas for the preceding year's R&D. Also, tell them that this is their only chance to request R&D money, so prepare their cases or hold their peace till next year.

Ask them to present their ideas with detailed designs, timelines, price tags, and effort. They should be able to provide an achievable end state and how it will get done.

Once the month ends, get the team together and ask them to present their cases to the group. It needs to stand up to peer review and not just attempt to fleece the boss for some "toy."

At the end of these presentations, get the group to rank the ideas in order of most beneficial and then see which of them your budget can cover. If there are arguments, just decide yourself and end the further discussion. A small tip here, should there be an argument, try to pick the project that rewards someone who deserves it. Tech people like to tinker with new technologies, and R&D projects are a great way to keep techs happy in their workplace.

After all this, you should know what you are spending for the year. You can keep the lights on, provide training and have some cool R&D projects underway. Of course, the major projects are also being funded and in-flight. Just make sure, as the manager, you know where all your money is going.

Cost Saving

Let us earn a couple of gold stars here. And by gold stars, I mean a way for you to add real value to your position and hopefully earn you a bonus or pay rise. Yep, in the first year.

While I mentioned that your CapEx budget should always be spent, there is no reason you cannot try to save money on your OpEx. Or better yet, divert OpEx into CapEx spending, maybe even for R&D projects explained earlier.

Saving on the OpEx is a great way to solidify your position as a valuable manager in the company.

I already mentioned that you are going through the OpEx with a fine-tooth comb and working out what is being spent on what. A big part of this is finding expenditure on anything unnecessary or over the top.

I have experienced many contracts where the company I worked for was wasting money on a service or system that was no longer needed. For some reason, my predecessor was paying for some cloud service or "test" hosting platform that should have been decommissioned many months ago yet was still drawing from company coffers.

Get onto these straight away. Close out what needs to be closed out and cancel contracts that waste company money.

Gold star one!

To earn your second gold star, look critically at each expense, from the largest down to the smallest.

A little time and research into these may create huge cost savings. I have found cost efficiencies on WAN circuits, wasted cloud compute, unnecessary licenses, and more. You just need to dig deeper than reviewing a line item and saying, "Well, it's there, so It must be needed."

Meet with the vendors. Introduce yourself to account managers and start pressuring them for a better deal. They commonly push their margins as the market is competitive, and their goals may have changed to keep their clients. They do not normally pass on savings to the customer until it is requested.

Take the time to review your needs, lower the cost of these and earn that second gold star.

When presenting your case for a pay rise or bonus, these savings are the tangible items you can provide your boss. "Look, I saved the company $X monthly and $X over the next five years." Powerful stuff!

Write Good Job Descriptions

You should have a good idea of what is happening within the organization. You should have started building a good repour with your team and have a deep understanding of where they fit within that team.

You should have created a Skillset matrix and know where the gaps are and what is covered. The next step is to look at the work, who is responsible for what, and make them accountable.

Let's start with the differences between responsibility and accountability. This is important.

Responsibility is the collective ownership of a particular system. This means a group can be responsible for a particular system or service. The key word here is that it is a group. Not a single person but a group.

Next is accountability. Accountability is when a single person is ultimately accountable for a service or solution. While a group of people are responsible, a single person is accountable.

So here is a scenario. The accounts team and you may be responsible for licensing a particular system. But you should be the accountable person for that license. If the license expires at any stage, yes, you are all responsible for it, but in the end, you are to blame for letting it lapse.

Accountability is a powerful tool for ensuring the continuation of services. Without accountability, issues will crop up a lot, and no one will take up the reigns to resolve them. Usually, the "that's not my job" mentality kicks in or the "that's what's-its problem." Unacceptable right? You do not want to be running around looking for someone to take ownership of a particular problem. This is where Job Descriptions come into play.

I harp on accountabilities and responsibilities because this is the primary purpose of having Job Descriptions.

You may very well have a team of IT guys hired to "manage systems," "level something support," or some other arbitrary term people use to generalize the IT support of the organization. Chances are people were hired under these loose terms and sit in a grey area of accountability. If nothing is clearly defined for your staff, they can hide behind it and shrug when push comes to shove.

Not only is this a difficult arrangement to manage it is also a way for the underachievers to do the bare minimum. You need to get on top of this quick smart. So, spend some time writing job descriptions.

Now you want to take your skillset matrix and prepare a new document. One for each of your positions or each of your people.

The JD may cover an individual employee. If your team is a lot bigger, it may cover entire teams, but it needs to cover what they are accountable for and responsible for. Let's dig into it.

Firstly, title your document the title of the position.

Next, at a high level, explain in very general terms what this position is to do for the day-to-day running of whatever they are running.

Then go a little deeper and explain in dot points the position responsibilities and how they tie in with others in that area of responsibility. This section can be copied and pasted across multiple positions, so put in the effort and make them clear. Remember, they are responsibilities. Thinking about what I said earlier about the differences between responsible and accountable.

Lastly and most importantly, dot out the accountabilities. Use dot points and ensure they can be ticked with confidence. Terms like - Accountable for the daily backup schedule, Accountable for the firewall policies, accountable for database maintenance. Stuff like that. No wiggle room. It is your job to do this, and you must own it. That is how a job description should be written.

It needs to be clear to the staff member that they are responsible for a bunch of stuff and share it with others but are accountable for critical areas of the business and must know what that is.

Once you have written out the JDs, review them with HR and make sure they update their end of things and approve of what was changed. They will file this detail away and probably appreciate your effort to make their life easier. Remember, HR does not know much about your area, so giving them clear and concise JDs for your team will only help. They shouldn't have any issues with the team leader clearing up people's responsibilities and should approve your changes easily.

Once you have the HR approval and perhaps your manager's approval, send these to your team and chat with them about what they are now 100% accountable for.

You need to be clear here. There can be no more grey areas, and you can even get the guys to date and sign so they understand what has changed (if anything). It may be something as simple as a storage engineer is now accountable for backups, but unless they acknowledge it, this task may stay with a collective and get missed.

Now that everyone knows what they are accountable and responsible for, you should have no further arguments, and everyone should be clear on their job. A massive step forward here and a good initiative to sort out quickly.

As time goes on and your team grows, shrinks, moves, or changes, you will need to keep returning to the job descriptions and updating them. Put a reminder in your calendar to review these once or twice a year. Refer to these job descriptions when you are reviewing team performance. You can focus on a member's accountabilities and determine if they are accomplishing the required accountabilities.

Technologies also change, and you find your team has evolved into looking after newer cutting-edge systems. You will probably find that the high achievers in your team will take these on because they want to learn. If the resource picking up this accountability makes sense, adjust the relevant job description to include this new technology. Don't forget, though, that you want more than a single member across new technology and will need to ensure the knowledge is shared.

Setup a Backup Strategy

Well, you are starting to nail a lot of what you need to ensure business continuity and working through many of your department's major functions.

An important part of your role is to have a disaster recovery strategy. A big part of that strategy is ensuring your backups work properly.

Disasters in IT are near impossible to avoid. It is a clear fact that incidents occur, and what makes a good manager separate from the average is a solid plan to quickly resolve these incidents. Having a way to recover systems that fail or become corrupted is an absolute necessity.

Let us start with a couple of terms that you need to understand. RTOs and RPOs.

Look at the diagram below. For those listening, picture a timeline running across a page. In the middle of the timeline, you will have a flash where disaster strikes. To the left of the flash will be an arrow, the length of which indicates your RPO. To the right of the flash is an arrow indicating your RTO. A simple yet powerful diagram is for anyone who wants to understand a good disaster recovery strategy.

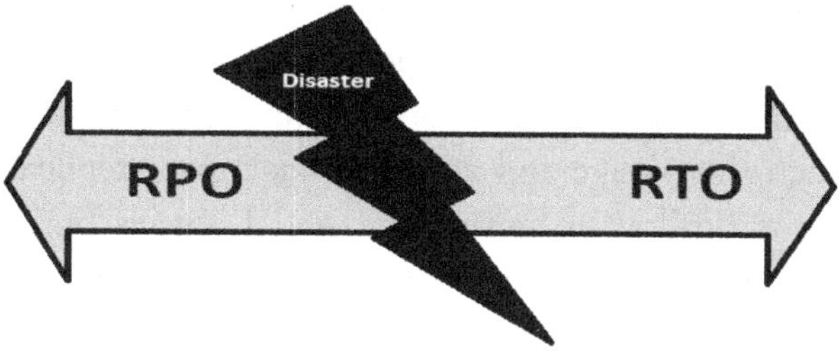

RTOs are Recovery Time Objectives. This is a time limit you set for your systems to recover from an outage. Basically, you ask, "How long is it acceptable for an outage to last." You should go ahead and ask the business this question and get an accurate answer. You do not want to accept a "Zero" as a response. If you have that type of business management, you must teach them realistic expectations. Failures happen, and you will need to deal with them realistically. How long is it acceptable for the business to be without key systems?

The amount of time they give you is important because, like anything, quick RTOs are achievable but will cost more. So, if the business says Zero in answer to how long systems can stay down and stand by that, you need to devise a strategy to deal with that. The cost is reduced if the business is OK with major systems taking a few hours. And if the business is OK with a 24-hour outage, the cost is reduced again. Just remember, it is not your job to dictate this. It is a business decision. It is, however, your job to inform them of the risks, costs, and expectations around your strategy.

OK, so we know now how long the business is OK with an outage. The next job is to work out the RPO. A Recovery Point Objective is how much data is missing between the outage and your last backup or replication job. It is difficult to get across, and I always put it in this context. How often do I need to snapshot the live business data? If I run nightly backups at 10 pm and I have a major outage at 9 pm, is it OK for the business to lose 23 hours' worth of data? This is another question you ask the business, and again, the narrower the gap between snapshots, the more the costs will increase.

When dealing with your recovery plan, these are two factors that need to be understood. As I said, determining what is acceptable here is not your job. It is your job to advise the business on its options and let them choose a reasonable one. With storage becoming cheaper and cloud storage more accessible, presenting a good recovery strategy that meets business needs is easier.

Once you have the business plan in mind, you can raise a project to implement the recovery plan and get it done quickly. If the business has agreed to this, you have made a promise. So, honor the promise you made. You do not want to be told to go ahead with a disaster recovery plan that includes a quick RTO and takes six months to implement it. Within those six months, a disaster could occur, your promise will be broken, your credibility will be shot, and the company may lose faith in your commitment. So, clarify the plan, seek the funding, and get it in as a priority.

Organise Outage Notifications

Set yourself up for success by creating a good communications strategy. A big part of this is to ensure your users are informed of what will impact them personally.

Believe it or not, I have worked in numerous places that think letting their users know a server is being rebooted is not important. I feel that this mentality is why users lose trust in their IT department. If IT deems it not important enough to let users know they are going to lose productivity for a period, how can they expect to have respect reciprocated?

So, set up a good process for outages. Ensure your staff communicates their intent for outages with enough time for users to prepare.

I aim for the one-week, one-day, 1-hour rule. That is, a week before a scheduled outage, send out your first notification. A day prior, send a refresher, and an hour before, send it one last time.

The notifications need to be standardized. I have used OneNote and email templates to do this, but it is up to you. The important pieces of information are the day and time, the reason, and most importantly, what the users need to do.

An example would be a major patch on your mail server. Let the users know when you expect to install the patch. Tell them why you are patching it and that their email will not flow during that period. You need to let them know that shutting down their email client is required and not to panic.

My template does this by answering a series of questions.

- Title for the outage description
- When is this happening?
- Why is this happening?
- What should users do to prepare?

Nice and easy. But especially important. You instantly gain a gold star from your users, and they appreciate the information.

Get your team to start sending these out before any outages. It gets them thinking about impacts on users sooner and gets them communicating with the users more often. Be part of the company you work in and not the cause of so many problems. People get that IT systems need maintenance, but they don't like it happening when they are busy with a critical task.

Side Note: Consistent Denial

Sometimes you will find that your outage notifications get responses that question the timing or will overlap with critical business needs. Most of the time, these are valid responses, and we need to adjust our outage window. This is fine, but what should you do if this occurs repeatedly and you cannot run your maintenance or change? This scenario is rare, but it can be a real pain to deal with.

You must remember that there are various reasons you are being told the outage cannot occur and your team is being told no. Generally, it's because someone places your team's work as less valuable than theirs. This is where you will need to step in as the manager and discuss directly with those involved. I find it best to talk directly to those responsible for the denial and maybe talk to their manager. You see, by telling your team that an outage cannot occur, they are saying that you are interrupting their work. This may be true, so you must find some common ground.

Ensuring everyone understands that this change or outage is an important task and must occur within a certain timeframe usually ends the discussion but, more importantly, gives your users a chance to come up with a valid time. You will need to step in, find some common ground and honor it by changing your outages to suit.

Nail the Asset Management

Time to start organizing the assets under your control.

Every IT manager should know what assets are within their jurisdiction. Having deep knowledge of the hardware and software you look after is worth taking time to gather. You will come up to periods in your tenure where you need to identify specific items quickly or pull up warranty details for failed systems. Having a database of this is important; it is your job to cover it.

Maybe the company has an asset management system that is easy to use, intuitive, and robust. Or maybe it's like most places, hard to navigate and has many incorrect or incomplete entries. You may even find the company doesn't have an asset management system at all.

I am afraid this is where you roll your sleeves up and take stock of what you should own.

It is not sexy work, but you need to do it. Or at least get your guys to do it. This is what you will need.

Everything you pay for, paid for, are responsible for, should be responsible for, or maintain should be listed in some manner. You need the following information in your list, and it needs to be accurate.

Column 1

Device, service, or license name – This will just be the general name you give to the item.

Column 2

Make – Exact manufacturer name

Column 3

Model – Exact Model name and/or number

Column 4

SKU – The stock-keeping unit of the item. Every IT asset should have one of these. You will need it to renew certain licenses or hardware if you need it in the future.

Column 5

Serial number – obvious

Column 5

Asset Tag (if existing) – If your organization tags items with its asset management system, record it.

Column 6

Purchase Date – The date the item was purchased or the license was activated.

Column 7

Market Value – Current market value of the item. Not the purchase price.

Column 8

Warranty Type – Next business day, 24/7 or on-premises. Make a note of the type or place a hyperlink to it. You need to know and understand the support you receive if this item fails.

Column 9

Warranty Expiry – What date will this warranty expire

Column 10

Yearly cost – What does it cost you to maintain this item each fiscal year?

Column 11

Service providing – What system does this item support?

Column 12

Current Location – Where is it? You can break this down by site, the floor, and the room if needed as long as you know where it is at a glance.

Column 13

Expected EOL – This is the expected end-of-life date of this item. IT hardware gets superseded quickly, but you should decide when this gear will expire.

Once you have this list, run an audit to ensure it is accurate. Try not to miss anything and regularly audit and update this list.

When making new purchases, place them on this list, and when you destroy or decommission gear, remove the items. I like to keep a record of what was decommissioned, also. You never know when you may want that data, and keeping a record of decommissioned gear with a decommissioning date is good bookkeeping.

Whether the organization does them or not, you should regularly audit this list against the actual equipment and have your team assist with it. I aim to conduct mine once a year, and some of my industry friends will run it every six months. The important part here is that you run it.

You should not be losing gear and should not allow warranties to expire. Expired warranties are an unnecessary risk. I have had multiple occasions where I let warranties lapse and found myself answering to management over why it took so long to repair a failure. Having to tell them a warranty expired is like saying I do not know what I am doing… Trust me. Make a big deal about gear missing a warranty and sort it out. You shouldn't be accepting that kind of risk.

Manage the Licenses and Support Contracts

Another key aspect of your role and something to grasp is the licenses within your organization.

Some licenses cover your support for systems, and some just cover the system's function. You need to take the time to understand the licenses themselves and what they provide.

So, to get started on this, you need to return to your asset list. You can add to this list or create another list but make sure you deal with each item.

The data you need to provide is if a license is required, when it was purchased, its price, and, very importantly, when it expires. What the license covers can be split into two factors. Function only or Function with full support.

What I mean by this is that you may have a license for a device just to activate the device itself and provide no support at all. Or, you have a license that covers both. You must understand which license does which.

Particularly important. Many companies sell devices with a license that just activates the device without providing any support, requiring additional payment. You need to grasp this as it can affect the resolution time on faults.

I believe any device you have that is not supported under a warranty and a support agreement is a huge risk. Your team may say the support is not required as the skillset is in-house, but that can change, and having a support arrangement should not be a choice but a mandate.

So, you should have noted all the licenses associated with your infrastructure. You should be getting a grasp of what the licenses cover and what to expect regarding cost.

Next, you should list the provider for the licenses and support agreements. Get in touch with each provider and request that you be the main contact for expiry updates and messages relating to them.

Add to this list your externally generated SSL certificates and Domain names. These aren't exactly licenses, but they have a term of expiry. You need to make sure these are covered in your OpEx expenses.

After all this work, you should have a definitive list of licenses and what they cover you for. Their expiry dates and maybe even their values. You should have contacted all the suppliers and added you as a contact for expiry info, and you should know where you have gaps in support.

Once you have this info and dealt with the gaps in support, you can fill out your calendar with your own alerts. You cannot always rely on the suppliers to get back to you about license expiry; your team may or may not pick it up. So, it will be up to you to mark your calendar with an alert to ensure you do not let anything expire.

Now, depending on how long your procurement department takes to pay for these renewals, you must determine the best reminder times for your alerts. A good rule of thumb is three months before expiry. Generally, that is plenty of time for you to organize payment on the renewals. If your procurement department is much slower or you want to revise the licenses each time through a lifecycle process, go ahead and give yourself more time. Or, if you can purchase within a day or two, adjust accordingly. You get what I am saying here. Give yourself enough time to organize payment on renewals.

Your document should be in your tool bag now, and make sure you keep it up to date with new licenses or support agreements as they come in. Ensure you keep it up to date, or it becomes worthless the first time you forget to update it.

Last point on both the asset list and the licenses. Take the time to clarify that these are in your OpEx budget. Ensure a line item covers each one and that the expense is managed. If you find licenses outside of your budget, look to have that cost reassigned to you. Take accountability for these costs and manage them properly.

Side Note: Revenue Generation via license renewal

Later in the book, we will discuss business growth, one of which is the ability to track customer licenses. This is relevant to those that are going to be managing a customer base, not just in-house IT systems. Using the tips in this section for your customers, you can generate great recurring revenue and track it for your business.

By ensuring you keep tabs on licenses as they are due to expire, automate a quote for the renewal process and sell them. I advise ensuring your customer licenses and contracts are tracked exactly as you would track your internal assets. Even if it doesn't end up a sale, it's always a great conversation to have with your customers about improvements.

Your checklist for the first month

Wow, what a productive first month in your new organization. You should have started setting yourself up for a quality future by introducing or, at least, getting a grasp of some of the key objectives of your role.

You may have been busy dealing with the day-to-day running of the team, putting out fires, and covering the organization's expectations of you. That is fine as long as you spend some of your time dealing with the items in this chapter. They are critical to doing a good job. You should be well on the path to a stable and organized IT department. Putting out fires all day will soon be a thing of the past.

So, let us check off the items you should have covered.

- Identifying and understanding your budget.
- Job descriptions and assigning accountabilities.

- Setting up a good disaster recovery strategy through RTOs and RPOs.
- Communicating change through quality outage notifications.
- Creating a record of assets.
- Getting a grip on licensing and support contracts.

The First Six Months

In the previous chapters, you have learned about the organization, how you fit in, and where you can get some quick wins on the board. You should have a good grasp of how the place works and implemented some good communications strategies. The next six months will be busy; as I have said, this is a guide. The items below may not be achievable in that time. I want to reiterate that do not let it stress you if it takes you longer. Some organizations are resistant to change. That is OK and perfectly normal. Just keep at it and implement these great initiatives if they do not exist. It will improve your team and the business. Some people need to be dragged kicking and screaming into the future, but once there, they are happy and respect your foresight.

We have a lot to get through, so let us dig into the next six months of your career.

Deal with the Unsupported Systems

In your investigations, you may have noticed some systems running the business that are both critical to someone and not supported by anyone. A good example of this may be an old access database someone created and has since left the company. The database runs on an incredibly old version of Microsoft Access, and no one knows how to update, change, or improve it. There are many more examples I can give you, but the basic fact is that these systems are important to someone, maybe even critical to the BAU of some department, and when it breaks, people will look to you and your team to fix it.

I have dealt with this in several ways, and the best advice I can offer you today is this. Do not sweep it under the rug. Do not pretend it is someone else's problem or hope it will disappear. The mentality that the system will just naturally phase out may be a huge mistake you make and will come to bite you in the future.

Take control of the situation. It is your job to both ensure these systems run and that they run well. Ignoring them will be a problem that can worsen as time passes.

Firstly, work out what the purpose of the system is. Ask those that use it what part of the business the system assists. Ask them what life would be like without it and what would happen if it broke. Even if it is one user, gauge the risk factors around this software or system.

Once you have worked out the criticality of the system and determined that it is impossible to phase out or decommission entirely, you know it needs to be managed and maintained.

The next step is to work out how to deal with it. Look through the skillset matrix you have created. Can anyone cover it? Have those discussions, and allocate the time to ensure it is covered. Add the system as a responsibility and make someone accountable.

If it has nothing to do with skills and everything to do with a much-needed upgrade or migration process, work out the costs. Determine a budget for upgrading or migrating the system to a modern platform and make your business case. Or, if you control one, add this to your budget.

Putting a business case forward will at least highlight the risk of keeping this unsupported system online. The rest of the management team may be unaware of this system's importance. Putting everything before them with options for resolving the problem is an important first step. I have had situations where I have informed the management team of ancient relics running the business and had a mix of responses. Some wanted the system decommissioned altogether and asked the user(s) to move on, and others have provided the budget to migrate to a cloud or other modern solution. And unfortunately, others took on the information and ignored it. Either way, the ball is no longer in your court. Those that choose to ignore have taken on the risk, and as long as you have everything well documented, you have done your job.

In the best-case scenario, this unsupported system is finally being looked at seriously, and the IT Manager (you) has taken action to ensure its longevity.

Set up a Ticketing System

Unless you have been living under a rock, you have probably heard of the acronym – ITIL. It has come in a few iterations, and when I sat the course and exam, I felt like it could be interpreted in several ways. I saw that what is supposed to be a framework was more of a guide.

Still, a few tools came out of my further research into ITIL, and I found that some were incredibly important and will make your life easier.

The first one I want to deal with in this chapter is having a ticketing system or a way to begin capturing important information centrally. Let me tell you that it is invaluable to have a ticketing system. If your team works without one, you probably notice how messy it is when someone goes on leave. You also notice how time-consuming it can be to detail work accomplishments for a previous period. You may even notice the same issues cropping up repeatedly with the same troubleshooting loops. These are all issues that can be resolved with a ticketing system.

Some call it a service desk, some call it a CRM, and others call it a helpdesk. People will argue with me over the differences, but I don't care. You should have SOMETHING to log all the work your staff do.

The basics of the ticketing system are simple. You need a place for all your staff to log their work. They should have a place to keep work issues open, where the team can collaborate, and close work out with good resolution descriptions for future reference. Everyone in your team should be able to share information from a central place. To me, the ticketing system is the core of an IT department. The mortar in the bricks.

There are 100s of these systems out there. They do different things and have added features that some people may need. I will tell you the most important aspects and what to look for when buying/trialing.

1. Every one of your staff must have an individual login. No generic shared logins (should not do this for anything anyway, but that's another book). Every one of your guys should log in themselves. No ifs or buts. If Active Directory can be integrated, it will make life easier, but it is not essential. Just have your team log in under their own identities.

2. Your staff should not work on <u>ANYTHING</u>, nothing, zip, zero, or nada, unless they have a ticket open. Get your guys into the habit of opening a ticket before they start work or pick up the phone. If no ticket is relevant to their work, they must create one. You can hold them to this as a KPI, and I will explain KPIs in a later chapter.

3. The ticketing system must have a great search function. This is where the value comes in. It will save you trouble later if your staff can find issues within seconds. It must be easy to use and intuitive, which means, for example, when you search an IP address for a customer, the system compiles the results in a way that shows issues to do with that subnet or customer's subnet. I have seen completely useless search functions, and you should not entertain a system like that. Spend some time checking search results in a live system to see what value that system will have.

4. It must be easy to create tickets. While closing a ticket may take time and effort because you want it to be accurate, the opening should be simple. One click on the main page, type a title and some context. Hit save. That's it! The ticket should save with the technician's name, date, and time. They can assign it to a customer during the closing along with resolution and other metadata, but the opening should be easy.

Remember these tickets need to exist first and be easy to create by your staff and preferably by your users.

5. Has a dashboard. This will be the most important screen in your office. You should be able to place the ticketing system dashboard up in the office somewhere for all to see and give everyone a presentable view of your world. If it is done right, it will be easy to see the productivity of your time by the sheer amount of open and closed tickets for the day. It should present unresolved, resolved, time to resolution, and perhaps a productivity graph for aesthetics. This is a real winner for you, so make it happen.
6. Your team can set priorities. Later in the book, I explain how prioritizing your tickets will be important to achieving SLA metrics and goals. Your ticketing system must be able to assign priorities to individual tickets to at least three levels. Priorities 1, 2, and 3. This is a minimum requirement. Some systems call the priorities different names, like high/medium/low. That's OK as long as it is translatable to prioritizing them.

You may get a lot of flak from your staff about having to create tickets, especially when they have not worked this way before. This should not be negotiable. Tie the tickets to their KPIs if you must. Just make sure the tickets are being opened, resolved, and closed.

There are multiple benefits to doing this. First is the fact that you can report on the work your team is doing. If you are pulled into a meeting each week, month, or quarter to explain yourself, show the management team your ticketing system report. Your system may even allow you to categorize issues, and you can talk about the top 5 issues your team deals with and how you will prevent them. Remember, your job is to provide direction. If you can see what is coming or have historical data, your life is much easier.

The second benefit is your team can share knowledge. You will always have technicians in your team that know a particular technology or customer better than anyone else on the team. They may solely deal with that customer/userbase, and that customer/userbase may only deal with that technician. This is OK. I do not have a problem with it apart from the fact that this technician is a human. They may get sick, go on leave, retire, or leave the company. Shocking, I know. When this happens, I have seen teams ignore the problem or respond with, "Ahh, they'll be back in a couple of days. They can get to it then". You should not let this occur. A client or user needs support, and it is your job to ensure that support happens promptly as they are used to. So, again, in comes the ticketing system to save the day. Chances are the customer's issue is already recorded in the system, and your team can read through some notes to resolve it.

Thirdly, a ticket system will allow you to measure your service delivery. Measuring performance and effort is incredibly important, and you cannot do this without a method of recording the data. The amount of tickets vs. time spent closing them is a good one, or the number of tickets closed per area is another. Trust me, having the tickets in the first place opens a world of possibilities for measuring success.

Create KPIs

In your first month, you wrote some job descriptions for your team. In those job descriptions, you wrote what your team is responsible for and what individuals are accountable for. This is an important task to set yourself up for this section, where you will start writing up Key Performance Indicators or KPIs.

I'll explain if you have never experienced KPIs in their many forms. Key performance indicators are your way to measure your team's performance several times a year. You provide your team with goals and expected results and then measure how these goals were achieved. It sets the stage for the year and allows your team to understand what is expected of them.

A lot of organizations have online systems for KPIs. They generally tie into the HR department and are reviewed by multiple management teams. Well, they should be. If there is no system, create one for yourself and your team. It can be a Word file you create with the year and interval for the meeting, but file it away securely and mention to HR or the equivalent that this is what you are doing.

I take KPIs very seriously. These provide direction for your staff and allow you to sit with each team member and discuss their progress one-on-one. You and your team members get to learn a lot from these and set your expectations for the coming year. Let us go through the process in detail.

Firstly, you need to create a document for each member individually. Enter their name into the document, their position, and then their accountabilities.

Secondly, set some goals for the year. Aim for ones that are attainable for them and the company. A good one is attending training for the year or ensuring company standards are adhered to. Three or four of these manager-type goals are fine. They can generally be generic ones that each of your team should attain. Do not put anything in here that can put undue pressure on your team. Asking them to pass multiple exams within the year is asking them to commit to after-hours study. While every technician or engineer should study on their own time, this is not something you should mandate. Remember, there may be family men and women in your team. Studying after hours to pass an exam may be too difficult, so keep your goals achievable while improving the individual.

Remember the ticketing system we set up earlier? One of the staff KPIs MUST be a percentage of productive time accrued in tickets. A good starter is 70% of work time. Therefore, in a 40-hour work week, you ask your staff to ensure 28 hours are within tickets. As the years go by, you can increase this number by intervals of 10%. You could measure this easily at each review period if you purchased the right ticketing system. Otherwise, you need to spend the time to ensure it is happening.

So, once you have set your goals for the staff, you need to sit down with them. Assign an hour for each person; the following is how I run my one-on-one meetings.

1. Run the start of year meeting in January of each year.
2. Start by explaining what you are trying to achieve by having the meeting.
3. Explain each of the KPIs you have created and that you expect these to be met before the year's completion.
4. Ask the employee to create 3 KPIs for themselves and have them back to you within five working days.

5. Ask if there are any concerns or anything they want to voice now they are having a one-on-one?
6. Voice any concerns about them that you need to have improved.
7. Ask if they need anything from you for the year. Now is their opportune time to ask.
8. Note everything said on the form.
9. Thank them for their time and say you look forward to seeing the 3 KPIs they will achieve.
10. Summarise everything that was said and email it to them. Ensure you clearly state what was promised and what is to be achieved.

The key to KPIs is they should be measurable. I have seen a lot of KPIs that are just fluff talk that really does not mean anything and can be explained away with more fluff.

KPIs need to be good quality and able to be measured. You should be able to put a percentage against each one, indicating its completeness throughout the year.

A good example is the training. At the start of the year, a KPI should be, let us say, completion of a Cyber security course. Your team member should be able to identify the course they will complete, which is entered into the KPI. Mid-year, when you discuss the progress, they should be able to tell you to have far through they are, and you can place that as a percentage completion against that KPI.

The poor alternative is something like "improve understanding of the job responsibilities through continuous improvement," as you can see, total fluff that could be anything or be explained in myriad ways. You must avoid this to show progression. Otherwise, the KPI reviews end up a waste of everyone's time.

Clarify and Renew Warranties

Always remember what the purpose of your job as an IT manager is. You must maintain and ensure the uptime of the systems in your care. But systems fail. You need to prepare yourself for this eventuality.

I would advise you to get in control of the warranties supporting the equipment you purchase or currently own.

You have taken stock of the gear in the previous section and worked out what they are worth to you and the company. Now back them up with a legitimate plan in case of failure.

Every bit of kit you look after should be considered valuable, and I think there is almost zero value in a piece of kit that is not backed by warranty. I always say that IT systems are as good as the warranty that protect them. I have had conversations about comparing technologies and how the price of one versus another does not make sense. They do the same thing. Why would we spend 20% more on this one? The conversations always end with the fact that a good quality warranty and exceptional support almost always back the more expensive. Take Dell over Super Micro, for example. In a cost comparison, Super Micro looks great, and you can get a great server for a low cost. The problem is when the parts fail; you are on your own with the Super Micro, whereas Dell will send a technician to help deploy and support failures, sometimes on the same day, to get everything up and running. Worth every penny.

Never purchase anything without a warranty, and never let these warranties lapse. It's a piece of mind; it's good practice and common sense.

So, in saying all that, go through your asset list, highlight the hardware not under warranty, and start renewing or extending them.

The difference between renewing and extending is self-explanatory; however, the extending part can be quite costly. For instance, you may have gone through your asset list and worked out there is a server that has just past its expiry. You can reach out to the vendor and have it extended, and based on your relationship, you may get a good price for another year or two, depending on the age of the asset.

The items that have lapsed for months or years may be a little trickier. The vendor may push you to refresh the asset, that is, to buy a new one. They may sell you an extended warranty, but it could cost you a fortune. You will need to weigh the costs vs. the benefits of doing this, but at least you are dealing with it.

Get busy with putting all your assets under the protection a warranty provides. It will save you from embarrassment in the long run and ensures you sleep at night. Trust me.

Take the time to call each vendor and work out what they will do for you in a failure. Start ticking off your list of assets and ensure they all have something backing them up in case of failure. If there is anything without one, see if you can purchase an extension or start looking at an upgrade.

The main point I am trying to bring across is that it's an important part of your job to make sure your kit can be replaced quickly should it fail.

Now, it may not be feasible at this stage to upgrade a piece of equipment that is out of warranty. That means you need to provide a warranty on it in-house. I do this by purchasing spares of that piece of equipment. Should it fail, you at least have critical spares in the same building that can replace it within the acceptable timeframe.

Either way, you must return the business online as quickly as possible without panicking and looking for a replacement part.

Keep this aspect of the book in your mind when you are purchasing. People may pressure you to buy cheaper gear that "on paper" can do the same job. You normally can stomp on this argument by looking deeper into the support and warranty of the equipment. Usually, the cheap stuff is not backed by any kind of warranty or local support, and you realize the purchase was a ticking time bomb. Avoid this scenario and give yourself a piece of mind by going for the industry-leading manufacturers with local presence and quality next business day (NBD) spares.

Build Relationships with Vendors

In a previous chapter, I mentioned that relationships with vendors are important. You may have reached out to them already for better service or pricing. That's good, but I want to clarify that you need to be able to rely on your partners and vendor throughout your career. These relationships may last you a long time, and even when you move from job to job, these contacts will benefit you for life.

Now that you thoroughly understand your assets, it's time to build relationships with the vendors that sell them to you. Whether it is your internal assets that will need to be renewed yearly or assets that you sell to your customer base, you must have a good standing with your vendors.

You already created a list of the assets, understand where they sit in the business and even worked out how critical these are.

It's time to reach out and get to know the companies selling them to you.

Most of your assets are probably sold to you by massive companies that turn over millions or even billions a year. You might be thinking, I am too small for them to worry about me. My company only buys a small amount of gear etc. You are wrong.

You are never too small to build a relationship with these companies. Sure, you may be introduced to the lowest levels of account management, but at least you know where you stand. The important aspect of this process is to understand that.

Reaching out to these resellers or vendors is important to understand whether they will or won't support you when needed.

I'll give you an example.

For some time, I worked for an MSP that was spending significant money through a reseller providing IT solutions for our customer base. We were not large by any means, but the spend with the reseller was healthy and recurring. I did not bother to follow the process in this section, and we went about our business. One unfortunate incident occurred where we missed peripherals for a solution, and we had no margin to cover it. It was a small amount of money in the big scheme of things, so I contacted the reseller to see if they could help. Not only did they flat-out decline our request the process taught me that we meant nothing to them. All the money we spent with this reseller was irrelevant, and they couldn't care less about our problem. Even with the amounts we spent each year, this supplier didn't have an account manager assigned us. We were nobodies in their system. I was incredibly frustrated by the whole process and was determined to understand how all our suppliers and vendors perceived us.

On the flip side, I had another situation with a vendor that truly cared about our business. Not only did they cover the missed items (I am talking smaller cost items, not entire servers here), but they expedited the delivery to ensure our project was a success.

By reaching out to each of your approved vendors and chatting with the account managers or partners, you will learn about how your company is perceived. Some are grateful for your business and happy to buy you a coffee to discuss the future, while others may be unable to find you in their databases.

As the IT manager, understanding and gathering this knowledge is especially important. Sometimes you need to lean on your suppliers for a quick turnaround. Sometimes they want to offer you deals that will benefit both sides. They may even fit into your roadmap for future upgrades and look to discount early for you. Knowing these people and befriending them professionally has a wide range of benefits. Get busy meeting them now, and you will benefit in the longer term.

Personally, I have taken these contacts with me from job to job and scored goals with my new employer. If there is one thing to take away from this section, it's this. IT managers are about leadership and providing direction but also about building relationships with people inside and outside your company. Vendors, suppliers, and partners should know you as the face of the IT team, and you should know them and how they can help you succeed.

If you find a supplier you are not happy with, be it they don't care about your business, are low value, or are just rude, take the steps necessary to replace them. There are a lot of vendors out there that will want your business and treat you with the respect you deserve. Do your research, go out for plenty of coffee meetings, let them sell to you a little, and build mutual respect. You want to help each other succeed in this business. Work with the people that will do that.

Implement a Product Lifecycle Strategy

Here is a term that gets thrown around a lot in this industry. You can quickly get overwhelmed by the information on the internet about lifecycle management, but I will simplify it for you as much as I can.

Product lifecycles or lifecycle management quantifies when a system needs to be retired or renewed.

Simply put, when you implement a system, piece of hardware, or application, you cannot move on to the next project or piece of work until you have noted when you want to revisit this system. These items will eventually have a license expiry, warranty expiry, EOL marker, or decommission date.

Let us look at an easy example. You have implemented a new service desk system for your helpdesk staff. Let's say it's ServiceNow. ServiceNow costs you a lot of money and requires a yearly license renewal. The common practice is implementing the system, getting the helpdesk team trained, and moving on. But I want you to think about a lifecycle for ServiceNow.

As the license is to expire after 12 months, set yourself a marker at nine months and plan a meeting with the key users of the system. You will sit down with these guys and determine if the system is achieving everything you wanted and is effective enough. At the end of the meeting, you want to know if you are sticking with ServiceNow, looking for alternatives, or increasing your

ServiceNow footprint. You should have an answer to these questions and a product direction. This is all lifecycle management is about. Continual review and continual improvement.

Now let us say in your meeting you find that ServiceNow has no traction and is costing the business phenomenal amounts of money with little to no use. Users are unwilling to pick it up, and the rollout failed. You dig deeper and find that a key component of the product should have been identified during the product evaluation.

Time to decide whether you want to reach out to the vendor to resolve this, ignore the problem and use the features available or admit a mistake was made and cancel the product. Without a proper lifecycle plan, you would never have identified this, and the company would be wasting money on a useless product. In addition, you would be chewing up your budget on a useless product.

So, for you, take out your assets list and set up a lifecycle plan for every item there. You can group assets like workstations or prioritize the bigger, more expensive items. Still, the important aspect of this is you are revisiting the solutions, discussing their value, and evaluating their need.

Develop a Training Plan

Next, I want you to focus on creating a good development plan for your team by ensuring they are continually learning. One of the unique aspects of our industry is it evolves incredibly fast. New technologies are introduced each year, which can revolutionize how our companies operate. Cybersecurity is an ever-changing landscape, networking continually changes, and our systems are continually updated and improved. It goes without saying that to be an effective IT team, you need to stay on top of the developments being introduced and ensure your team can deploy and maintain these systems.

Not only do you need to stay on top of these to provide efficiencies or improvements to the business, but your team will also appreciate that their manager cares about their development. Remember that developing your staff is one of the key factors to retaining them, if not the biggest.

Now you may work in an MSP that requires a certain level of certification to keep a partnership, or you work for a company that likes to keep up with the latest. Either way, a development plan is essential to move with the times.

A development plan is quite simple but usually implemented poorly. Some organizations expect their staff to learn or improve in their own time, and some do not understand what working in this industry is like. That is where you come in. Time to sort this out as a matter of importance. You want your team to continually learn, and you need your company to understand this is part of the norm for a quality IT team.

Let's dig in on where to begin.

You created a skills matrix and determined your team's strengths and weaknesses. This is essential to moving forward. From there, you have identified the gaps in the knowledge or at least where there is no overlap between members. Start with these.

Determine how you will train your team to fill those gaps and provide overlapping skill sets to cover core technologies. You do this by talking to the individual members, whether it be during the KPI reviews or casual one-on-one meetings. Ask them to identify what course they would sit to cover the gap, the cost, and the time required to complete it. Store this information as it is vital to the development plan.

Next, work out the weaker skill sets within the team that need to be covered. Same again, get them to take the time to identify the course, costs, and time required. Put these into the development plan as well.

Your development plan should start looking something like this;

Member	Gap	Course	Institute	Cost	Time	Schedule
J.Doe	Networking	Cisco CCNA beginners	Didata	$ 5,700.00	5 days	TBA
J.Bloggs	SQL	Advanced SQL techniques	HandsOn	$ 3,200.00	2 days	TBA

The schedule can be TBA for now, as you are just developing a plan.

Once you have completed your development plan, the next challenge is to execute it.

This can be a major challenge as, up until now, your company may have been ignoring training altogether. You may be approaching your budget or the CFO and worrying about how you will cover the training costs. Well, you need to. One way or the other, you will need to keep your guys trained and learning constantly.

The best way is to give your guys five days a year to leave the office, sit on a course amongst their peers and learn the subject. There are alternatives to this, depending on the outcome of your discussions with management. So first, raise the subject of a training plan with them, see if there is a mutual understanding that training is required for IT, and come to an arrangement, hopefully, a budget increase.

If you are flat-out denied an increase or people balk at the costs, try an alternative approach I have had success with. The main issue with instructor-led courses outside of the office is usually the high cost. 3 to 5 thousand per team member can quickly add up, and even after all the savings you provided earlier, management may see the cost as a waste. There is an alternative.

Once you have exhausted every avenue for a solid training budget, you can look to option 2. Online training.

If you have never seen one of these, I would be surprised. There are many online training systems, some good ones specific to the IT industry. Pluralsight is my favorite, but you might like CBT Nuggets or Udemy. There are many more, each with unique offerings, but the important aspect is the content itself. It needs to be quality, and it needs to cover your training plan.

This is our plan B. If you have hit walls trying to obtain a training budget for your team to go offsite, this option will require a smaller expense and some time away from the desk.

You should be able to get this one approved. Sometimes, the licenses are a few hundred dollars, providing your team access to every course within the chosen online training system.

Whether you can achieve offsite learning or have your online training licenses, you have the means to activate the plan.

All that is left is to work out the schedules. This is an important commitment when you can afford the time away for your team. Many places think organizing online training is fine and expect it will be done in personal time. Change that perception. Make it so proper training is conducted away from the office within the time allocated. That is why the classroom is ideal. It gives your guys a week to learn and chat about technology with their peers. The online training still needs time away from the desk and not after hours. So, organize it properly and officially and measure what was learned or completed. Remember that you may provide time away for your staff only to find out they didn't even bother to log into the system or attend the course. It's unlikely, but it can happen. Ensure that whatever you choose has a way to measure productivity or attendance.

Lastly, update your skills matrix based on the training conducted. Your Helpdesk guy may have learned a new skill, which should be added to the skills matrix. Keep tabs on your team's growth holistically and work year to year to strengthen and improve the skillsets across the team.

Setting a development plan for your team is vital to the organization's future and your staff's retention. Seek funding for this critical initiative, choose a system that fits the budget, and then start scheduling in your team. Measure the improvement and update your skill set matrix.

Address the Single Points of Failure

Now that you have a good overview of your assets and the systems you look after, you will need to investigate a term everyone in this industry should know – Single Points of Failure.

If you do not know what this means, it is a term used to describe a technology that is tagged as critical and, if a failure occurs, will bring down critical services until it is fixed.

For example, your head office may have all centralized server infrastructure hosted in the head office and a whole bunch of remote sites connecting to the head office to gain access to shared resources like POS systems or email. The head office has a single internet connection that your predecessor was proud of because it's fast or "business-grade."

That internet connection is a single point of failure. A major one. Should that internet connection fail, all your remote sites will go down, and your head office staff will be unable to connect. Many services may fail until the internet is brought back online, and should you have a Service level agreement (SLA) on that link, the business will lose productivity. This should be unacceptable to you.

So even though that connection is fast and sounds great, it loses all its value the first time it dies. You need to avoid these types of single points of failure. It is your job to take note of these and remediate as many as possible.

Take the time to go through your systems with a fine-tooth comb and ask what happens if this fails? What is the impact on the business, and worst case, how long until it is brought online? There is a common process in the industry called Disaster recovery planning, and this section covers part of a good DRP. Where DRP might be a book all its own and creating DRP is highly involved, I wanted to cover the very basics of single points of failure, which should lead you into heightened critical thinking. So, let's focus on single points of failure for now.

Note all the items you have addressed with the question, "How bad will it be if this fails?". They need to be collated into your initial assessment of the current state of the systems.

Once you have identified all these problems, start working on a way to resolve them. In the example, I gave earlier, a secondary, albeit slower, connection on a separate carrier is a small price to pay to mitigate that risk. Push comes to shove. You can divert your traffic over the second link and call the carrier to speed it up. Problem solved, and you are the hero because the business keeps running.

So, go through the list of single points of failure, remark on how big a risk it is to the business, and then place your recommendation for mitigating them. Collate all of this with the costs involved in fixing them and send involved your management chain.

This is what you are here for. You need to think ahead and be proactive. There are a huge number of IT professionals and managers out there that will pull out the excuse book when failures occur. Still, only a few have the foresight to do this work and put in the hard yards to convince the business to take it seriously.

Yes, it is your job to convince management of the seriousness of your proposals. It is easy to mention or whinge about this to your peers. Anyone can do that, and trust me, many people do. These people are usually the dregs of our industry. If you find yourself having a whinge with one of your subordinates, STOP. You should never gripe down, always up. If you want a whinge, do it with your boss, not those below you. This will just harm morale and never helps your standing in their eyes. My motto is that losers whinge, and winners do. Your true value is shown when you act on the issues that can affect serious productivity, not just complain about it.

So, when you send this up the chain, make sure you put plenty of effort into clarifying the risk, discovering the best solution, articulating the solution well, and backing it up with costs and experience. Be prepared to create a business case to back yourself up and keep up communication with your management to see if they are being followed up. I will talk about business cases later in the book. This is a vital skill for any IT manager, and having the ability in your toolkit is important.

Best case, your proposal is approved, funding is given, and a project can be raised. Worst case, you are ignored but no longer accountable. Meaning should the items you brought up fail – you can explain that this item was uncovered previously, and you have a permanent fix ready to go.

Now you should be well on the way to ensuring stability and resilience are introduced where they were missing. You have researched solutions to single points of failure and communicated them to the management. Your value to the business increases even further.

Build Resilience

In the previous section, you identified and addressed the single points of failure. Hopefully, you have covered all the Single points of failure in your organization, and now you can look at revisiting these regularly.

Maybe the term resilience sounds like a buzz word and in some circles, it may very well be, but this should be important to you.

The systems you are accountable for keeping online must remain online for as long as possible, and resilience is the best term for it. Like my points on single points of failure, you should consider the following points to determine your resilience.

What resilience means is an absence of single points of failure. Making sure that what you have online stays online in the event of a catastrophe or failure event. Making sure there are redundant paths into your Datacentres. Having replication of data wherever you can. Having a network topology that allows for a switch or router to die.

This section differs from the previous one because you are starting to look at systems regularly and continually auditing them against risk.

You need to spend time, year after year, taking an extremely critical look at your systems. You have taken the time to audit them and, by now, know where everything sits in the big scheme of things. Now you must enable your "Continuation of Business" (COB) mindset and start thinking about worst-case scenarios.

I do this by asking, "What happens if this dies ."Simple question, and if you hear the words "we lose everything," you need to note that. What you are making notes of is a single point of failure. If a single piece of IT infrastructure fails and you lose everything, you are at risk. You don't want to lose everything ever. Losing a single service can be accepted as the rest of the business can work around that service while you get it back online, but losing everything is simply unacceptable.

I also go through and ask about previous major outages in the company. The team may remember a memorable event that caused major grief for the IT team and the company. Any major outages need to go into this list you are creating.

So, note the items that cause a "lose everything" or "lose most" scenario and then take the SME or accountable engineer aside and work on a strategy to mitigate those risks.

Sit down with them and go over all the options and scenarios.

Maybe you can use existing equipment in place to provide redundancy, or maybe you need to spend a fortune on making a redundancy. Either way, you should set a strategy to ensure the systems are resilient.

Don't let the spending a fortune aspect of this worry you. We will discuss presentations later in the book and the best way to nail the funding, but you need to work on your strategy first.

You do this by following an amazingly simple workflow. Review and improve, rinse, and repeat. There are loads of workflows that include additional steps, but at the core are these two simple steps. I follow them to ensure what I am doing is relevant and beneficial.

As I said, there are workflows with multiple steps to improving your IT resilience and many models that can be studied and introduced, but I want you to start here. Take this simple two-step repeatable workflow and introduce it to your organization.

Let's start with the review process. Well, you have already done that in the previous section. You identified and addressed all your single points of failure. You have also reviewed your staff's skills, capabilities, and many other aspects to identify your standing. These are all aspects of building resiliency. Lock these in as a regular occurrence. You should constantly review what you are doing and take a step back to ask yourself if it's working, working enough, or needs change.

That's the second part of the simple workflow. Improve.

Now that you are taking the right steps to identify your usual processes or tasks, you must continually improve them. In this step, you are spending the time to see if these items can be improved.

For instance, your single points of failure are being addressed. You implement a solution to resolve the single point of failure. Give it some time, review if that implementation worked, and can it be further improved even if it did?

Here is a good example of the workflow in practice.

Earlier in the book, we mentioned a single internet connection to your head office was deemed a single point of failure. Your resolution could have been to deploy a secondary internet connection for fail-over purposes. You deployed it. That's great. Maybe even tested that it works. But what if, during your next review, you find a way to load balance these connections, maybe even reduce the overall costs of your WAN connectivity? What if, during your next review, your network engineer mentions an even better way to deploy through some new technology he or she discovered? Should you not continually review these items, this would not even occur to you.

So, now the tangible stuff. I want you to look back at your budget. There are multiple items there, and you should understand what they all are. Add a column or create a new spreadsheet where you identify a timeline for reviewing each one. It may be every three years or every month but put this in your calendar and involve the right people. Spend time reviewing these items and look to always improve on all of them. This is the path to building a resilient team, department, infrastructure, and, overall, a resilient business.

Your Checklist for the First 6 Months

Such a busy half year. Keeping busy is a good thing, and I hope the items up until now have been both challenging and rewarding for you. I also hope you are finding it fun. Sure, it's your job, but you should enjoy at least parts of it. You may be living in hell at this point, and it's all too much. Maybe you can evaluate whether this sort of work is suited to you. Remember what I said initially about this being a different type of work from your typical IT functions. Managing people and processes can be difficult.

Much of what we covered in this 6-month plan may be new to your organization, but it can only improve how the IT department is being run. Stick to it.

As I said before, you may not get all of this done in the time frames I give, but this is perfectly normal and nothing to worry about. Keep at it and get to this point of the book before moving on, even if it takes you the remainder of the year.

Let's see what is to be checked off.

- Identifying and dealing with those unsupported systems
- Setting up a ticketing system

- Creating KPIs for your team
- Vendor relationship building
- Creating product lifecycles
- Developing training plans
- identification and resolve single points of failure.
- Building resilience through a simple 2-step workflow

Side note – If you are finding that you have not scratched the surface here, I want you to pause at this point in the book. As I keep saying, these deadlines are not set in stone, but you should have partially started and carried on with a few of these initiatives. However, if you find it difficult or struggle to manage your time, I want you to search this book for time management. I talk at some length there about managing your time better so you can focus on these important initiatives. So, if you are struggling to get most of these past six months crossed off, fast forward to the time management section of the book. Understand what I say there, and then come back to this section.

The First Year

You are on your way to creating a solid foundation for your team and company. You are covering some important areas here that may have never been considered previously, and I would like to think you are kicking goals for yourself and your company.

This is just the beginning. You still have a lot of work to do, and I am sure you are filling up your calendar nicely and communicating your changes properly, as I keep saying. The timing in this book can be a loose guide. If you have not gotten to this point in the book within six months, that is not a reflection of your incompetence. It is merely showing the need for your efforts. Maybe your company is not agile enough to take on change as it should. All is well. Even if you are into year two at this stage, let's dig into the next phase and tie up some loose ends that may already be going through your mind.

Implement Product Standardisation

The equipment and technologies your organization use for itself or its customers may vary and be challenging to maintain. Each engineer in your team may feel one product is better than another, or every salesman in your company sells what they know. There is nothing wrong with people having a preference, but supporting this type of mentality is incredibly difficult.

On countless occasions, I have seen solutions deployed, and they end up supported by a sole engineer in the company. The reason they chose the solution was for this or that purpose, but no one else wants to touch it. Other team members either don't have the required skill or feel it was a mistake to deploy in the first place. Because of these reasons, they are hesitant and may even be hostile toward the solution.

As I said in a previous chapter, team members might be long gone, and the solutions they chose end up being unsupported over time. This scenario must be avoided. You need to maintain a standard, and if there isn't one, let us get involved and set a standard for your team to follow.

Finding the best solutions can be daunting, and a lot of research is required to do this right, so let us start with what we already know.

Skills of the team – what equipment or technologies can your team support? This is a good starting place. We have a matrix we created previously, so dig that up and start with that.

What partnerships are in play? – does the business currently have any arrangements with vendors or suppliers that unlock good pricing? You should have reached out to the vendors by now and discussed the benefits or lack thereof of dealing with them. Understanding these will help you choose the best standards for your team, your company, and your customers.

What is used currently? – is there an Ecosystem already being worked on? Does the business lean towards any specific vendor type? If so, why? This is also completed previously. You created your comprehensive asset list and have asset management in play. Using this information will help you determine if there is a trend, a standard, or if we are looking at a mess of different solutions that create a headache.

Now let us look at what we may not know.

What are your customers buying? - Are your customers after cheap, basic, or enterprise-grade, reliable, top-shelf systems? Maybe they sit somewhere in between?

This may take some time and research on your behalf. Going through previous sales or touching base with your sales team is a good start. Take the time to understand what type of customers you sell to and what they are likely to buy. Your sales team is focused on the future, but knowing you have their back and are preparing better customer service is a great way to have them on your side. They will need to understand your coming changes in strategy and that you will standardize the solutions being delivered. They will want to understand the end game and what you are proposing as they will be pushing it. So, get to know them, their customers, and what they need to achieve success. A standardized solution is a great way to align with them and ensure clarity for proposals and quotes.

Once this research is done, gather your skills matrix, asset list, and sales team/customer information. This information should do to begin your list. Now take stock of the technologies you are selling or supporting. Make an exhaustive list, and do not leave anything out. We are talking from desktops to servers, firewalls to modems, Internet services to wireless systems. Make that list and include everything you are selling or supporting.

Once you have your list, make two columns and call one silver and the other gold. You can always go for a bronze/platinum if necessary, but we have two standard options for now.

Now that you have your list and want to fill out the solutions, talk to your team. Sit the group down and start going through each solution group – Firewalls, switches, servers, etc. You will find a lot of great insights into preferences for tech. You may even see some relief on the team's faces as this may be the first time they are looking at standardizing on hardware.

First, keep "silver" the cheaper option and "gold" the expensive one, and make your list with your team. It is a fun exercise, and you may learn a few things.

Make sure you book plenty of time with the team. I would organize a day for this strategy session and lunch. This process will take some time and is a pivotal moment for your team and the company. You must clarify to your team that once this list is created, any purchases or proposals outside this list will be unacceptable and must go through an approval process. Purchases made outside the standard solutions have a high chance of rejection, so discussion needs to be robust, and agreement needs to be democratic. Some team members will be resistant or have an issue with the requirement to standardize or don't understand your policy. Try to explain the benefits as I have summarised previously. If that doesn't work, you must use your influence and explain that standardizing is mandatory for your IT department. Their input is critical to its success. If you still have resistance from certain members, continue without them but make it clear - purchases outside the standards will be rejected without solid business cases for bypassing the IT product standards.

Upon completion of your product standards, send them out to everyone involved in the business's solutions, designing, and procurement portions. Make it clear that these are agreed upon by the business and make it clear that acquisition outside of the standards is against company policy.

Create Compliance Documentation

You may find yourself in a position where you must provide direction on Cyber security items. You may be in the position I have found myself in and are getting various information from various sources. Not so long ago, I was told that the security of our system is where it needs to be and, at the same time, told the security of the same system is weak and vulnerable. People see the systems differently, and you can become confused when finding the correct answer. What is the best way to begin, and what should you focus on?

Take it from me. Start by finding yourself a compliance document relevant to your industry and read it from start to finish. These documents are written by people who are experts in the field of Cyber Security and write documents to ensure security measures are in place to suit your industry. There are standards now for everything from Government entities, construction to banking. Choose the right standards for you, and go find the documents. They are readily available on the internet.

In the scenario where I was given information from various sources about right or wrong, the correct compliance document put all the arguments to rest. I quickly found that you have the necessary answers and details in these compliance documents.

There are several good documents out there, and depending on what part of the world you are in, a good start is a Google search for "(your industry) Cyber compliance document."

Look for documents with clear control guidelines. A good example of a control will be something along the lines of "Firewalls must control traffic between your protected and unclassified network portions." Controls like that are gold because they clearly define what MUST happen in your network. There are terms like <u>must</u> and <u>should</u>, which could confuse you. "Must" means that a control absolutely must be implemented. The term "should" is more of an option, an option that you should go for if possible. If it's impossible, mitigate the control some way through the process or by locking it down in other ways. A good practice is to take the document relevant to you and highlight the controls you must use.

Once you find a document or two clearly stating the requirements to comply, you have a great way to focus on your future state.

The next step is to compile the documents into a single document for your organization. Take it from me. It looks good when you place the org logo at the top of your document, title it "OrgName ICT technical reference," and paste in the relevant controls.

Make sure your controls are worded so that they can be answered as "Yes, we are compliant" or "No – We are not compliant" There should be nothing in between. If you have these controls worded poorly and there is an in-between, then you need to reword it. It must be a yes or no.

Here is an example to get you started.

Network Access Controls

"Administrators must have unique credentials when accessing network systems for administrative purposes."

Simple yes or no for this control. Yes, they log in via Radius or similar for accessing switches. No, if they all use a generic local account. By the way, it's also a no if there is a combination. Make sure your answer covers the whole field of the control, not just parts.

Write up a draft of this document. Take your direct manager through it so they understand what you are trying to do and have it approved. Put all the approvals into a field within the document as "approved by" and save it as version 1.0.

The next step is to pass it down to your team so they understand what the future state will look like. They should all take the time to read through it and understand that you have a clear direction for the systems.

While they are taking the time to read through your document, some may not, but that is OK. Having a professional IT governance document passed down to them may be new to them and a little daunting. They just need to know that changes may be coming their way and to prepare for an audit.

So yes, while they are taking the time to read through the document, you can get started on an audit.

I will take you through auditing later in the book but make sure you are ready to put your money where your mouth is on these controls. A lot of IT controls may mean less flexibility for your end users. Removing social media, for instance, may cause a major backlash, so you need to prepare for this and have buy-in from senior management to assist with the coming complaints.

Stay the course though. You will undermine your credibility if you present a strong security-focused posture and then renege because someone in accounts does not like it. You need to be prepared to stand by each of your new controls and protect your team from the whinging that is coming. And it will come, so be prepared for it.

A memorable one for me was a particular control we implemented to remove USB access for users. It is an important control as it limits the opportunity for viruses to enter your network. Without expensive sanitization systems, it was our only means of protecting ourselves from the unmanaged data people place on USB sticks.

Obviously, this was met with many arguments like: "But, I need to work on documents from home" or "Customers provide me with valuable data via USB."

In most cases, this is untrue or embellished. You must work around these and find solutions to problems that do not weaken your security posture. Again, stay the course.

Most people don't like change, but I will tell you this. The organization you work for may not be able to survive change, but they are almost definitely unable to survive a security leak or Crypto locker attack.

There is a long list of companies that have gone under or fined large amounts of money due to lax cybersecurity. Do not be one of them. Protect what you are responsible for and stand up for what is right regarding security. It is part of your job. When the proverbial hits the fan, your neck will be on the chopping block alongside the user who introduced the compromise.

Anyway, I can carry on about the battle between functionality and security, but regarding this chapter, remember that a large number of compliance documents are out there to help you create your network and systems designs. Use them. You may be smarter than most people, but experts in this field write these docs. Listen to what they have to say. Write your ICT compliance document and stand by it.

Enough about why compliance is important. Let's look at the next steps.

So, you have put together a list of controls for your organization. You have worded them so they are yes or no, and you have explained them to your organization and team.

Time to take the organization through an audit. I will take you through the steps for an audit later in this book, but at this point, it is a good idea to go through an initial audit. That's right. I am asking you to skip over to the chapter on auditing now. While I wanted a nice flow to your tasks, this step must happen now. At least to gather the initial data on the company's compliance standing.

Auditing sounds like a real headache, I know. The term itself is seen as a policing of poor practices or even snooping around. In some ways, it is, but it does not have to be a painful process. Skip down now to auditing and take a first run at it. Come back here once you have completed an audit of the organization.

Develop a Roadmap

Let me start this chapter by saying road mapping is something many IT managers or businesses don't consider. At all.

It always frustrates me to join a thriving business that seems to make good money and has an idea of how to drive revenue growth but no roadmap in place for the future. Especially as the act of putting a roadmap together is not hard.

Nobody likes to request emergency funds to fix a major problem within a business, and no business likes to spend money that was not forecast. Of course, some businesses have emergency funds for issues that occur but imagine if you can reduce or eliminate that emergency fund purely by forecasting the future changes that need to be made.

I have also found that the response is always positive once you put together something, anything, to roadmap future technology for the IT systems.

So, this chapter is all about that. Now we have already discussed product lifecycles. You know, the need to continually refresh hardware for failure avoidance. I hope you have assembled something and have an asset register to monitor your warranties, licenses, and EOL systems. If you have not done any of that, go back and learn that chapter and make those moves. If you have all of that squared away… good work. This all gets placed on the roadmap.

The roadmap is all about identifying key milestones that need major expenditure and resources to deploy them. It is that simple. So, you need to start looking at things for the long game. I start with 5-year roadmaps, but ten years is not hard once you get your head around it.

A good roadmap is a timeline that highlights major projects that your team is working on. It should present to anyone picking it up what is happening in your world and when. It is a graphical representation of key objectives over a period.

When putting your roadmap together, the key point is to first look at the technology being replaced within the chosen period. Enter those as part of your roadmap first. If you will refresh your storage array within the period, add that in. Storage upgrades are a major project.

Secondly, enter major projects you see as relevant to the business that also fit into the budget. Work with your team on what will benefit the business most and add it to your roadmap.

Do not forget your compliance requirements from the previous chapter. If you are following along, you should have a set of controls that are either being met or need to be mitigated. These should be part of your roadmap. Lock these items in.

Here is a good example. Say your roadmap has space in year 3. Let's say you will have $2M to spend. The business you work for has an incredibly outdated security system. A system that is currently problematic and constantly on your team to resolve, using up valuable time and energy.

Well, there you go. Modernise and simplify that system. Make it easier to manage, and spend the 2 million on a brand-new security system for the business. Enter that onto your roadmap.

It may not be as simple as that; you may need to put up a business case. But that is good. Having a business case means it is not out of the question and just needs justification. It means the roadmap is working, and people are taking it seriously.

Continue down that path. Fill out a roadmap from year to year until you cover the next five years and have it ready to present.

You may notice a couple of things, firstly, that you will not have the resources required to complete all the projects within the time allocated. This may be a justification for growing your team. Secondly, you may notice that your budget may not cover one of the years within the roadmap. You can adjust the timeline accordingly or prioritize other work to achieve your goals. Adjust everything so it makes sense to you and you have the right knowledge to ask what you need from your organization. Then you are ready to present your thorough roadmap.

In the tips section of this book, I will discuss how to deal with presentations, so have a read of that to get your head straight and get your peers and upper management to review your roadmap and approve it. This is crucial. Keep at them to ensure your roadmap is approved at the highest levels. Once you have that in writing, your next five years are mapped out. It is just on you to deliver after that, and when you are looking to spend company funds, it is not a surprise to anyone. Well, the accounts team always acts surprised, but that is their job…

Anyway, here is an example of a roadmap that you can replicate. A picture paints a thousand words, so once approved, stick it up on a wall for the team to see what the future looks like.

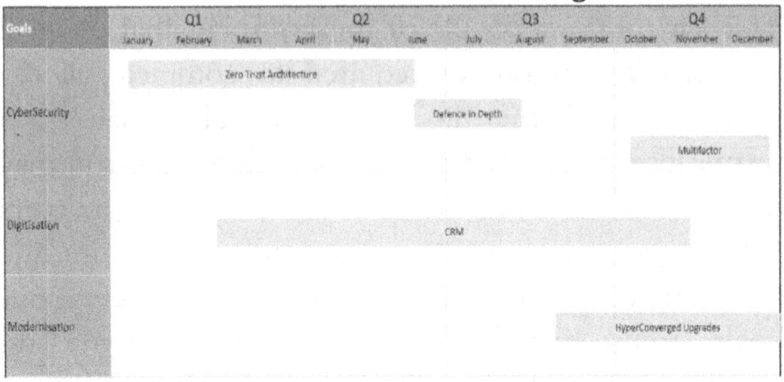

You should put together a few words explaining each step along with your roadmap. There are many ways to format the document, be creative or search the internet but spell it out for anyone reading it. The diagram you create paints a great picture but not everyone understands the three-letter acronyms of the IT world.

From experience, creating a roadmap like this one works in three stages.

Create a first-run draft of your roadmap, and try to align it to your budget as best you can. Keep your descriptions at a high level and tag each initiative with a budgetary price. Do not spend too great a time nailing the description or the pricing. Just have enough to be able to talk senior management through it. This needs to be presented in a draft format to your boss and those able to justify the budgets in future spending.

Once you have the nod from those above to complete the roadmap, take it away, add more details, seek out vendors for better pricing, let them know when you anticipate the project kick-off, and continue updating the descriptions. You are now aiming for a published version of the document ready to release to the wider team.

Publish your document. Send it off to your boss and as high up as possible. Send it around to your team. Paste the image on the wall and get ready to take on the projects as they kick-off.

Remember, once you and your company have committed to this roadmap, it is up to you to deliver them. Ensure you will be ready for each stage, as your success will be measured by delivering each project on time and within budget. It's important that your roadmap is as close to the real timelines and cost as you can get.

Organise Change Management

Another aspect of your job and a key to ensuring that the IT department is not seen as a bunch of cowboys is to get your changes under control.

Many IT teams have the mentality that they are in control of when systems come up or down. Because they can do it, there is a mentality that they are justified in making system changes whenever it suits them or their timing.

You should not encourage this mentality, and you should clamp right down on it. The systems are there for the business to be more efficient at what they do. It's the users that run the company and benefit from these systems and not the guys managing the systems. Remember this. I'll reiterate because it's important. The users are the drivers for IT and not the IT department. The users are there to keep the company running in the right direction efficiently. Your job as the IT manager is to ensure these systems are always running when needed and make them efficient and as cost-effective as possible. This means you do not just bring down a production server whenever you want or run patches in the middle of the day. It also means that your network team does not test links by bringing them up and down to test redundancies. This is what cowboys do.

So, what do you do to manage changes to an IT environment? Change control.

ITIL provides you with the knowledge to deal with change, and there are many ITIL tools out there that will help with the actual process, but I will try to explain the most important parts of change control and how I have dealt with it.

First, you must understand that a change control process is about communication. It is about letting the right people know what will happen within and outside your team and giving them a chance to postpone or even stop it. It's about ensuring you have thought of everything before you make a change or bring down a service.

Secondly, it is about really thinking about the impact of what you are doing. Taking a step back or having your peers review what you want to do may illuminate an issue. Sure, it's not always fun having the rest of your team blow apart your work, but if you don't stand up to your peers' scrutiny, then chances are you haven't thought through what you are doing in the first place.

Thirdly, change control is about thinking through the worst-case scenario with the change you want to implement. While you may think a firmware update provided by the manufacturer will work fine, what if it fails a system? What if a function you rely on stops working after you roll out? What if it bricks the hardware? You must consider how you will deal with that and have prepared steps to roll back.

Let us talk about what aspects of a change need to be recorded. You may not have the expensive tools that deal with change control, but I will list what you should have as a bare minimum to ensure they are good for action.

Firstly, get your team to start thinking about the impact of the work they want to do. Will it cause an outage or a change to the production systems? Yes – well, that is a change request. Patch Tuesday can be considered a change, as some patches require servers to reboot. This is an outage and needs to have a change control in place. Changing a group policy to have a new background presented to desktops – probably not a change request, but it may as well go through the process anyway. Changes to the testing environment (if you have one) are probably not a change request, as this is where your guys should tinker. Update to a firewall to allow traffic into a server that is a change as it may impact existing traffic flows… You see where I am going with all this. Your team needs to think critically about their work and the impact that work can have on the production environment. They need to start thinking about the end users and that what they change may affect the end user's work. Keeping the users happy and productive should be in our very nature. It's what we are here to encourage. Not the whim of an IT guru who thinks the world revolves around them.

OK, I have explained what to focus on as a need to change control. What about the next step and setting up a process?

First, you have your team identifying the changes needed. It is good that they stop and think, but now they must request the change. It is no longer a sole decision. Annoying, right? No. What really will be annoying is the team responsible for stopping operations because of some random update no one knows about.

So, your first step is to explain that changes are no longer individual decisions. They will be going through a formal process, and you will email everyone the new process to be clear.

In the next step, your team should now seek to formally request a change. They must think carefully about what they want to do and compile it for a change meeting. I'll get into those meetings here too, but for now, these are the questions your team needs to answer before raising a request.

- What system am I changing?
- What is the end state of the change?
- Will there be an impact on the system being changed?
- Will there be an impact on any dependent systems?
- When do I want to make this change?
- Can the change be automated for after hours?
- What is the actual work to be conducted?
- Are there any users impacted? Who are they?
- When do you want to make the change?
- What is the time frame for an outage?
- What is the rollback strategy?

These changes should not be done on a whim. These questions stop your staff from thinking, "It'll be easy," and thinking more along the lines of "Have I covered all my bases."

However, if you get this information compiled, ensure it is done. Nothing should be done without these questions being answered and then discussed.

Previously I mentioned spot meetings at the start of each week. Well, now you must conduct a change control meeting at the end of each week. Set the schedule and add the agenda. Ensure that any change requests from the team are made a day before the meeting as a deadline. Change requests after that time will not be discussed.

Emergency changes aside, change control meetings gather all team members and discuss each change request. Make it formal. These are called CAB or change advisory boards in the industry. It doesn't matter what name you choose as long as you make it happen. Set the meeting so that only change requests that have been submitted are to be discussed and not an opportunity to raise one. Any missing the request deadline, which you set as the day prior, gets skipped to the following week. Discuss each one and allow the member who raised it to explain what they want to do and what the end state should be. Allow everyone at the meeting to have an opportunity to question the change and discover additional impacts. Should everything be above board, approve the change and hold the requester to the specified time. You ensure the guys have crossed the Ts and dotted the i's. There is no need to make this a hard process. Just get the team thinking about what they are doing and ensure minimal impact.

Changes need to happen, but they can be painless. If everyone knows about it, there shouldn't be any problems.

Part of the approval is not just a formal nod from the boss. You also need to ensure the guys send a formal outage notification. I have gone over outage notifications previously in the book, so ensure that the process is followed for these changes.

At the end of the change control meeting, everyone in your team knows about outages coming in the next week. Those that requested changes have the approval to go ahead and let the impacted users know about them in preparation for making them.

As a side note here, be very wary of two things. One is minimal effort or information put into the change request. You may have an engineer that just puts in the bare minimum information or explains at a high level. Throughout the meeting, always ask yourself whether the information is enough to clearly explain what will happen. The before and after states are clear to you? No? Reject it. Ask for further information. Nothing wrong with saying you need more information and that the change is too hard to understand or too high level.

The second aspect to be wary of is complexity. I am not saying that a change can't come in with difficult concepts; I am saying is the change segmented correctly. I will give you an example.

A change came across my desk for a "Network hardening" update. The details were all there, but as I read through it, I could see the changes were planned to be made across multiple devices within a short time frame and could see a nightmare waiting to happen. Within the change meeting, I asked the inevitable question, can this be broken up into smaller changes over a longer period? After some to-and-fro from the team, we discovered the engineer wanted to update too many things at once and that all had a lot of risks associated with it. A process over several weeks was identified very quickly, and multiple changes needed to be implemented. Luckily it was done this way because just one of the updates caused issues that needed to be rolled back. The change control process saved us a massive headache and a lot of downtime.

Back to the change control meeting. The final part of the meeting should be a debrief of the previous week's changes. Go through last week's change requests list and bring up each one quickly. Ask each member that was involved if there were any issues.

If there were issues, discuss them and work out how to ensure they do not happen again. It is important that lessons are learned on these issues, so take the time to share them and learn from them.

If you have a knowledge base, put these lessons learned into them, but at the very least, have them in your ticketing system.

Support Levels

In your career, you may be asked to create a Service Level Agreement (SLA). Or you may need to review SLAs through certain vendors or outsourcing.

SLAs are talking about our boundaries regarding supporting systems and services. Now you may have never had to deal with these types of contracts or heard it as a term used in higher-level meetings. Whether you have an understanding or not, I am going to reiterate what these are and how to set yourself up with an SLA. It's important to set these boundaries as you can measure against them and set expectations for the business you are hired to support.

Let's dive in.

You may be in a position where units within the business or the customer say they do or don't receive the support they would expect. You should ask yourself first what is a good level of support. Are the things we do to put out the fires enough, or should we be doing more? Can we even afford to do more? Does the customer expect us to do more? How did this expectation come to be?

You have just spent many months noting down everything you support and who owns what, so the next step is to assess the situation concerning levels of support.

Let's not make this any more complex than it should be. An SLA defines the responsibilities, how long you will take to react to a problem, and how you react. That's it.

So, take your customer or business units and make a note of each of them. Make a note of the key business line applications they use, and then work out how critical these applications are. Next, all you need to do is work out a manageable plan to react to outages.

Write these out clearly and concisely. These are the starting blocks of the SLA. They need to be written so that there are no grey areas about it. For example – Your organization looks after company X's ICT infrastructure. That infrastructure consists of A, B C systems. Clear and concise. This sets the foundation. Remember, having zero room for argument is what you are aiming for here. If you don't note something down, you may end up having to look after it, and that is why we are going through this process.

Once you have set out what systems, infrastructure, or solution is covered, it is time to work out support agreements. You have just clarified that you look after systems A, B, and/or C. Next is what an appropriate running state for those is. Let's say system A is a virtual server environment, a bunch of virtualization hosts, and a SAN. You clearly defined this in your document, so now you need to explain clearly and concisely what a running state looks like. Powered on, redundant power, online, CPU and RAM running levels, warranty, etc., need to be mentioned. You want to clarify everything that needs to be looked after and who is accountable for its running. This is the important part. You and your team may be accountable for the CPU and RAM but not for the power. So, you write that up. Company X (in this example) will provide power and power loads for the systems. Company X will provide fully redundant and stable power to the virtualization system.

My company will ensure that CPU and RAM are optimized and available for all virtual machines Company X is deploying up to a certain level you define.

I may be losing you here. That is OK. All you need to do for now is list all the systems that sit on or around what you support. Really dig into everything here. Take out your asset list if you need to, but in essence, break up your SLA document into parts and explain who looks after each one. Once you have determined the stuff you look after, the next step is to define how you will deal with issues.

Start by stating how you will acknowledge an issue.

- Who can your customer contact, and how?
- What timing can they expect acknowledgment?
- What timing can they expect support?
- What timing can they expect a resolution?

A good plan is generally to have an issue acknowledged within an hour. That can mean contact from your helpdesk and additional notes added to a ticket within an hour. Setting up an email address with an auto-response and a ticket being generated is a really good way to meet this SLA target. I have set this up a few times now, and most modern ticketing systems make it easy to configure this. helpdesk@mycompany.com is an easy setup. Depending on whether your team is next business day or 24/7, you can mention so in your response email. The customer at least understands when a human will take up the ticket and run with it.

From there, you can set a plan to tag severity to these tickets, and based on your SLA, you can determine a time to resolution (TTR) for them.

The table below is an example of what to add to your SLA. It needs to be clear and describe what you categorize as a priority 1, 2, or 3. Try to ensure this aligns with the capability of your ticketing system. Your table should also show how quickly your team can respond to each priority.

	Description	Response	TTR
Prio 1	Any incident that causes business-wide failure of a critical system. No workaround is available.	1 Hour	2 Hours
Prio 2	Any incident that causes manageable failure of a critical system. No workaround is available	1 Hour	4 Hours
Prio 3	Any incident that causes impact to an individual. Workaround available	1 Hour	NBD

You can set all issues in your organization to have a response time of less than one hour. This is easy to achieve as all that needs to happen is your helpdesk marks the ticket or the call as responded to. An automated system can notify the ticket creator, letting them know it is under review. SLA is met. Hold on, you think, you have not done anything to help that user! Not true. By placing the ticket into a system, you are in the first step of supporting them. From here, the magic happens, but at least the first SLA is met.

Next is your agreement with that business unit for time to resolution (TTR). This basically means that if you have agreed to resolve the issue in 2 hours, you can escalate or resolve the issue in a manageable time frame.

Priorities are important. You should have a goal to contain all issues into three priorities. Priority 1, 2, and 3 minimum. Priority 1 issues being the highest impact on the customer or business. Prio1 would be a massive issue that affects the whole business and does not have a workaround. Priority 2 is of medium impact. This is an incident where only a part of the business or a department is impacted. There may or may not be a workaround for it. Finally, priority 3 would be your usual weekly adds, moves, or changes. There is a workaround, and the impact on the business is minimal.

Here is an example. Let us say three tickets come in at once. A POS system failure, a username change, and a workstation slowness issue.

The POS system failure would be your priority 2 incident and may be under one of your SLAs to have a 4-hour time to resolution.

The username change is a priority 3 task that can be achieved the next business day.

The workstation slowness issue is logged by the CFO of the company, who has explicitly requested 1-hour turnaround on issues related to them. Maybe in your agreement, you agreed to ensure that any C-Suite users are immediately flagged priority 1.

No problem. Because you have the SLA in place, you can hang back on the username change, escalate the workstation issue to the desktop team as high priority and have the POS SME review the POS system ticket ASAP and within 4 hours.

With a decent ticketing system, these types of issues are easily manageable. You can see how it returns on your investment when you see it in action.

SLAs are great for managing those expectations. You can measure everything that is going on and write reports based on your SLA. At the end of each month or quarter, you can present your report and explain why you are missing SLA targets (lack of resources, funding, skills, etc.) or get a gold star for meeting your targets again, hopefully for the 5th, 6th, or 10th month in a row.

Sometimes SLAs will have penalties associated with them. If you are reviewing an SLA for your organization with an external vendor, you want to ensure these are realistic and appropriate. Say you are reviewing an SLA for external support on an application key to a business unit. You want to make sure the penalties are sufficient due to poor support.

Your company wears a loss of revenue due to a particular supported system going down, and if you are not at fault, why would you wear that loss? Protect yourself by making the penalties adequate.

Negotiating SLAs is also a good skill to have. You want to be part of these conversations, and there is no harm done by asking what you are expected to provide or asking the vendor what they will give you should their service fail you.

I have negotiated numerous SLAs around different services and managed to have some great protections added to our contracts. Sometimes your inquiries can be a real headache, but at the end of the day, your job is to protect yourself and your business. A new software package may look great for your company, but it is only as good as the support behind it. Don't forget that. You spend large amounts of money on a solution to find that it's run by one developer from his garage that will shrug when it fails. Not good enough.

The key to an SLA, and my point in earnest, is that you need to spend time reviewing your external vendors and the support they will provide you. You must ask, "Is it good enough for us, and does it align with our business."

A good example is one I have dealt with recently. The vendor, located and headquartered in Italy, provided a critical production system. Their SLA provided support during business hours… In Italy… Therefore, we could only ever communicate major issues with them during our evening, and my engineers were spending their nights working through problems. I had to give them time in lieu and acknowledge their work-life balance problems. Not a great SLA for us at all.

In the future, you can find issues like this, evaluate them, and potentially reduce the cost of support agreements. Failing that, you can look elsewhere. Simple as that. It's your organization's money, so why not get value for it?

Understand Salaries

Previously we discussed budgeting and everything you need to be across when it comes to the financials of your department. I hope you are even across how the organization's financials sit and where your budget fits into that equation. Good knowledge of and understanding of the company's profit and loss helps you become an effective manager. One of the key aspects and probably the largest expense is the staff salaries. You may have already investigated or dealt with these in your end-of-year / mid-year reviews. Either way, I want you to take the time now to delve into the salaries of your staff and understand how they are laid out.

The objective here is to understand the value the salaries provide to both the staff member and the value the staff member provides to the business. In a nutshell, you need to work out - does the salary fit the engineer? Is it above, below, or equal to the industry standard?

Disclaimer: The next step is extremely confidential. I assume you are good with maintaining confidentiality, but in case you are not. This new document you are about to create needs to be secured. House it on your internal server, password protect it, and ensure only you or maybe HR have access to it. Under no condition should this document ever be shared with your team members.

Now that's said, create a list of your staff and add their current roles next to each name. I like to start from the most junior to the most senior. Remember the job description portion of this book? The JDs will come in handy for this as they will clearly define each staff member's role. List them out, their titles, and then add their salaries next to that.

Now you need to go out and research. You may need to reach out to your recruiter and ask them for a coffee. Tell them what you are doing and that it is an activity to ensure you align with industry standards. You want to understand if your staff are at, above, or below the standard for their skills and experience.

There are also recruitment agencies that publish annual salary benchmarks that you can use. Here in Australia, a recruitment firm, Hays, will provide a nice list of the typical IT positions and salary ranges for different states. It is a great tool to use when you are determining a baseline for your team or if you are dealing with roles that are outside your experience.

Doing this type of research will help you understand the value of the engineers in your team. I'll give you an example. Let's say you have a guy on your team that maintains the helpdesk. He is competent and does a great job, but he never extends himself or up-skills in any way. You can safely lock him in as a helpdesk or service desk admin and work out the industry standard. Easy one. Review his salary. Is it on par? Great.

Let's try a more difficult member of the team. Your network engineer is an absolute gun. She nails the networking side and supports the storage team above and beyond what is expected. In fact, the interactions with that team are what you should consider as mentoring. Clearly, she is multiskilled, and your skill set matrix reflects that. But does the salary? How do you even quantify what she is worth to your organization? You can, and you must.

I always take people like her and ensure they are given 10% above the market average based on the skillset matrix being 10% above the requirement. It's a tricky concept, but let's take the network engineer as our example. She does her networking and should be on par in the industry as a network engineer. That's easy. But what about those storage engineering skills? Well, you can't pay her double. She won't spend the same amount of time dealing with the storage as she does with the network. She also won't work double the hours. But she must be rewarded for her skills in that space. So, this is where you work out the percentage of time she spends on the extra stuff.

Whoa, are you saying she is now being paid for multiple jobs? Not at all. She is hired as a network engineer first and foremost and is getting paid for it. What we are doing is clearing up the fact that she is competent in her role, killing it, in fact, and is able to fulfill the requirements of a network engineer and support other teams. This is an efficiency that is worth rewarding. Working out how much time is spent with the other teams is not too difficult. Provided she has updated tickets and the tickets reflect the time she has spent supporting the storage team. If this is unclear, just ask her.

None of this is something you reveal to your team, by the way. Soon you will have people knocking on your door, letting you know they are helping this team or that team and deserve a pay rise. Chances are it will be to the detriment of their daily duties, so keep this to yourself. I am trying to teach you how to deal with the superstars of your team and how to determine what a fair raise looks like, and what quantifiable value they bring to the team.

So, in the case of the superstar network engineer, all you do to understand her value is work out the extra effort spent on top of the daily duties and quantify that in a fair salary.

Now, back to your document. You have listed the salaries of each staff member as they currently are, listed whether they are above, below, or on par with industry standards. The last piece is to put the salary you expect to adjust. You only place this adjustment figure in if the member is below the industry standard or if you believe, based on the details we discussed earlier, they need to be increased. You want to highlight the members of staff here that need to be adjusted to suit your valuation of them. At the end of this process, you will see an overall view of who in your team is being paid fairly and who is below your valuation. This will help with your end-of-year reviews and help with deciding who is getting a larger salary increase.

You should review this sheet every year, keep it up to date with all new additions to the staff, and use it to evaluate future employees' salaries.

Your Checklist for the First Year

You should have had a very busy year. The items up to this point can take a significant amount of your time, and as a manager, you can start to see why it's no simple matter of looking after a team on top of everything else. All of this is in addition to what would be considered BAU tasks. It's a lot, and as I keep saying, don't stress too much if you cannot cover all these items by the end of your first year. This rings true, especially if your organization has not implemented these items prior to your tenure. If you are rolling a lot of this out from scratch, then be prepared to be at the end of your first year and only have dealt with a small portion of these items. Sometimes change takes time. Some organizations are resistant to change, and their new IT manager may be doing too much too quickly. No worries, just keep at it. All these tips are incredibly important to becoming a competent IT manager. Stick with it.

Let's review the checklist for the past year.

- Implementing product standards
- Creation of compliance documentation
- Development of a technology roadmap
- Implementing change management
- Creating and revising Service Level documents
- Understanding and reviewing employee salaries

Year Two and Beyond

Well, if you are up to this point, you have implemented some good initiatives. Your team and the IT systems must be performing well. Your team is squared away, and the business has a top-notch IT manager running a tight ship. You are probably impressed with yourself and have senior management respecting your every decision and leaning heavily on you for anything technology related or otherwise. Time to sit back and relax, right? Wrong.

There is still plenty to do, and the ever-spinning circle of lifecycle management continues. Sure, you can focus on improving all your implemented processes and ensure everything runs better and more efficiently. Still, I want you to focus on adding to your workload with some items that keep those bonuses coming in year after year.

The problem with sitting back and marveling at your genius is that the organization has raised its expectations of you, your team, and the IT department. They will have short memories. We all do. They are just going to reason that the status quo has been this way since they can remember, and by doing nothing, people will question what you even do around here. Well, not always, but it may start to eat away at you that nothing has changed recently. Why am I even here?

Honestly, I feel that way all the time. Being challenged is important. If you have the business up to this point, there is a good chance you are starting to feel like you got them where they need to be and on to the next challenge. It could be a good chance to move on and implement everything you have learned here in the next place. Plenty of them need these initiatives, and successful managers like you are a valued commodity. If not, you like the place and being amongst everything you helped build; then there is still plenty to do. Let's investigate what you can focus on over the next few years.

Auditing

So, you have your compliance document and are keen to implement it. What to do first?

Audit.

Yeah, it's a nasty word in many fields, and people neglect to dig around in their systems looking for issues, but you won't be that person. You should start to have a mindset that audits make you stronger. Finding mistakes is good because you learn from them and make your team and systems stronger, better protected, and more efficient – the list goes on.

In a previous chapter, we discussed creating a technical reference document where you created controls specific to your company. In the document, you listed many compliance controls and whether these controls are compliant - yes or no.

Let's improve this document by making an audit checklist. Take each control and give it a number. Next, copy and paste these numbers and their definitions into a spreadsheet. Once you have all your controls from 1 to 100 or more in your spreadsheet, you will need to number and categorize these for a future search or reference. Make column one your control number. Column 2 is the definition. Column 3 is a simple Yes or No. Column 4 is the reason. Column 5 is the risk posed. And column 6 will be the mitigation plan.

You can only fill out columns 1 and 2. The remaining four are for your audit. You're all set.

Now let your team and management know that you are running an audit of the ICT in compliance with the technical reference previously approved and endorsed by management. They are to be prepared for questions and investigation.

Next, go through each one of the control points and tick yes or no. Are you compliant, or are you not compliant? There is no in-between or grey area. You either do the control, or you don't do it. If you have a control that can be answered in any other way, you need to reword it and try again.

As you go through these controls and tick them off as compliant or not compliant, fill out the fourth column with a reason. Do what you can to fill out this column, ask the right questions, research, check the configuration, etc. but make sure you have something here. So, if your control is "two-factor authentication must be configured across network management systems," fill out column 4 with the system you have in place to provide this. Otherwise, tick no and respond with no system to provide 2FA. This whole process helps with understanding the systems in place. It's also a good practice to test these controls and make sure it's not just Schrodinger's measure. It must be a working solution. Otherwise, you are lying in your audit.

Finally, column five. This can be done when you return to your desk to finalize the audit. If a control is in place, then it's a "Not Applicable or N/A" If there is no control, you will need to detail the risk posed.

Once you have all this detail, you know your system's state of play. This report should be sent up to management and to your team to understand that there are gaps in the compliance of the systems and potential security risks.

You can now create a plan to start mitigating these risks or changing your systems to become compliant. You may need to raise many projects to manage these changes. I will take you briefly through the basics of project management later in this book. You may even add these mitigation actions to your roadmap. Hopefully, you can see how all of this is coming together.

Business Growth

Another focus point should be continued growth in your business or the company you work for. Now there are two sides to this, and if you are looking after internal IT systems, you would be somewhat excluded from thinking about growth. Still, if you are running professional services or managed services team, you need to start looking at how you want to grow your team.

You see, growth is the best measure of your success. Having more customers through the door and being able to hire more members to your team means you are winning. Sitting stagnant and dealing with the status quo is not the best representation of your talents. You have spent the past year or so squaring away your team and ensuring they can manage the workloads. It's time to improve further by showing the company you can handle more and increase profits.

I know what you may be thinking. This doesn't interest me, or I'm not a salesman. I get that you are entitled to think that way, and you can be assured that dealing with what you have is a noble cause. But if you want more, if you want to bring those bonuses in year after year, showing growth to your company can be a great way to earn that. Let me show you how.

Start by reaching out to your sales team. You may already have a great relationship with these people and have organized a good process for dealing with their requests, but I want you to reach out to them for a different purpose.

Have a chat with them about growth for your team. Set realistic targets. Say another member per year or a 10% increase in work per year. Something realistic.

Work through a plan on how you expect to achieve that target. Most of the time, the conversation will end up being educational, with you and the salesman learning about how you can support each other. Mention that as the head of the department, you are willing to help them with pitches to customers, design stuff, or even just regular meetings with the client themselves to put a face to a name. It all counts, trust me. Having a presence in front of the clients is always a good idea and keeps the sales team honest. Your company will be seen as a trusted advisor, not just a number cruncher looking to sell more products. Keep the cost of things separate from the conversation and let sales discuss that. Focus on the technology and how you can help the customer.

You will soon find yourself providing advice and squaring away deals with your sales team. Getting a little buzz each time they win a sale. Remember, every time your sales team wins a sale, no matter how big or small, that work will come to your team and get you closer to the goal you agreed to. The goal you set is to grow, and if you support sales in their process, you can do just that.

The Circle of Life

You have come a long way to get here. So many changes and initiatives are now introduced to your organization. They are lucky to have you onboard. If you have implemented all this, you are 99 percent of the way there. Honestly, the only improvements from here are up to you and your ability to dig even deeper. Our industry changes so much, and while I could go on about all the different managerial processes that can benefit further, I will leave it here. I find the state you are in after achieving the goals in this book is a great state to be in and one that cycles nicely. You will find that year to year, you will be dealing with great projects looking for growth opportunities, and continually improving through your audits and road mapping.

Let's recap the yearly tasks you should cycle through.

- Auditing and improving deployed systems
- Updating and delivering the roadmap

- Growing the business

Tricks of the Trade

We have come a long way with this guide and have some great focus areas as the months and years go by. In the next section, I wanted to provide some of my learning experiences. I mentioned a few items in this book that you might not be familiar with, like when I first tackled them. The next many tricks are a quick guide to some core skills you need to succeed as an IT Manager.

Time Management

It's quite easy for us to end up in a rut where we work 10 – 14-hour days or maybe even weekends. Unfortunately, sometimes this is unavoidable. This is the nature of our work. Following the steps in this book will limit the amount of after-hours work as you gain control. Sometimes difficult issues pop up, and you need to work through them, but this should not be the norm.

You might say, "I am working long hours even though everything is running smoothly," Or, "I need to work long hours to prove my value!"

…not true.

Firstly, you do not need to prove your worth through anything other than smooth running ICT operations and proper reporting. The rest is considered a change if systems are online and running to the requirements. Change has its process outside of this topic. I want you to consider this point. Urgency should be for systems outages ONLY. Everything else is a change request. Therefore, you start to see your emails and messages in a new light.

On a Friday, when the VP of finance emails "Urgent: need another distribution list created for some random idea of mine" directly to you, it may sound like something you need to deal with immediately. It may look like an action that requires a quick turnaround, but that DL will turn into more work, and that work can take you into the weekend. Usually, when it's all done and dusted, probably no one will need this change till a month later anyway…. This sort of stuff happens all the time. I'm trying to teach you that it's not that critical. That VP of finance can log a ticket, and it will be done.

You can revert to the SLAs you created in the previous section to support your stance. These SLAs will clearly state the effort required to deal with work coming in. Time management is the main reason you create them.

Remember, the rest is considered a change if you have systems online and working. Change can wait for the proper processes like the ones you are setting up in this book.

Secondly, the world will keep spinning if you do not respond to that email or message. Just because someone places "URGENT" in the title does not necessarily mean it is. If you start responding to every message in this fashion, you will find it will be expected of you. You will end up burning out and working ridiculous hours. In my eyes, you are not considered a hero or hard worker. You are just working stupidly. Sorry.

People will always have different ways of measuring your value. Some may think you are important to the business because you are there from 6 am to 6 pm. Some may think you provide value because you are good with customers. Some may even think you are valuable because you look good. Not kidding.

Don't worry too much about what people think. I truly believe this is where many people burn out or get workplace depression. These issues crop up because people worry too much about what others think. You can't control what people think; they will think what they want, so try not to stress about it. Worry about what you can control instead.

Your job is to keep the business running through the IT systems. Keep the business data safe and accessible. There may be tasks slightly outside of that but in a nutshell. This is it. Remember this when you start to feel disheartened or when someone takes a swipe at you. And most importantly, remember this when you feel like you need to work excessive hours.

Anyway, here is my tip for success in work/life balance. It works for me, and I hope you try it.

Every Monday, you should prepare for your week. Go through your communications, emails, etc., and determine their priorities. Over time this will get easier to do. For now, do your best to prioritize what needs to be first or last. Once you have your list of priorities, schedule via a calendar, assign time to each task for the week, set up meetings, and mentally prepare. You should have a plan for the week by lunchtime on Monday morning. Follow through by having your spot meetings with the team and even announcing your goals for the week. Get them to do likewise.

From Tuesday to Thursday, execute your plan. Any new high-priority stuff may come in, and that is fine. Just reschedule your workloads to deal with them and shuffle things around. The key is to work hard each day within your allotted 8 hours. Get it all done. You made a promise on Monday to yourself and your team to complete the tasks you allocated, so do it. Eight hours a day to nail meetings, reports, and updates. Work hard. Your goal is to knock off after those 8 hours feeling accomplished, so get to it.

I treat Fridays a little differently. For me, Friday is to clear out the work log for that week, so all those items you rescheduled or new items that popped out over the week are getting sorted on Friday. Any new items that come in on a Friday are left for the next week. That's right. If you get any emails on a Friday, avoid dealing with them unless it's urgent. Remember, we already talked about urgent work, and chances are the emails you get on Friday are not urgent work at all… So read them, digest what is being said, maybe seek clarity from the sender but ultimately, move back to your work log. The idea is you are clearing your plate as much as you can for the weekend.

Work hard on Fridays. Harder than any other day of the week. You are at the last leg of the week and want to nail as much work as possible to relax on the weekend. Weekends are for downtime and unwinding. Make it so you feel accomplished and you have earned it.

You may be like me and hate to leave things incomplete, so work hard and do as much as possible. The important aspect here is that you are done by the end of Friday. If systems are online and you have completed your workload for the week, go offline. I mean it. Shut down your machine, go home, and enjoy your life. You have done all you can. Monday is another day, and as I have said, the world keeps spinning whether you answer that email or not.

Timeboxing

A very close friend of mine introduced me to the concept of timeboxing. It's a great way to achieve goals and focus your efforts on the tasks that matter.

Timeboxing is the basic process of filling out your calendar with important tasks to help you achieve your goals.

We all know that there are multiple business tasks you need to achieve as a manager. Meeting management, your team, customers, and the like. These will take up most of your days. I get that. I want you to create time slots to focus on your long-term goals. I add them to my calendar as Focus time. Other than Monday, it's a slot of time every morning for a couple of hours that I allocate to dealing with the long-term goals I am trying to achieve.

I want you to do the same. I want you to slice out a couple of hours each day for "Focus Time." I want you to add the goals in this book to that recurring task in your calendar. I want you to strike out any you have completed and spend your Focus Time on these items, pushing them forward. The next piece is to ensure when someone wants to eat into your focus time, you defer them to another time in the day, say no, or agree that it cannot be postponed and eat into your focus time.

Remember that when you start compromising on your focus time, you are basically saying that noise is more important. Try to avoid this.

Delegation

In case you forgot, you are now the IT Manager. You are now responsible for a team as well as the business systems. I can't stress enough that focusing on the items in this book should be your highest priority to provide true value to your role. Dealing with items you may have done as an Engineer is not the best use of your time. You should look at the bigger picture; delegation is important to your efficiency. Delegation doesn't mean passing the buck and going off for golf, well, maybe it does to some people, but you're not like that. Delegation is about passing the work to the right person to handle it effectively with your guidance and support, should it be required.

As a leader of your team, you will be passed on various requests every day. You need to always think about whether it is relative to your role or can your team handle it. Trust me when I say that most of this new work is something your team can handle. You may want to handle it, but that shows you lack trust in your team's capabilities. Among other things, it shows you are not comfortable with passing the work around to those best suited. This is narcissistic behavior. Try to avoid it.

You can't expect to work with your team or have a close-knit group if you take everything on yourself and/or pass annoying items you don't like. It needs to be a clear decision process that shows you trust your team with almost anything related to their field.

If you are like me, I often throw my team into the deep end and see if they sink or swim. I am almost always pleasantly surprised that they not only swim but exceed my expectations. I know my team appreciates this approach as they feel they are trusted with challenging work.

Just make sure, whatever you want to delegate to your team, be clear that you are there if needed. When delegating work, remember to explain the expected outcome, provide time for clarification, and promise to support should it be needed.

Retention

Always a critical topic, especially in this industry. Retention challenges are something I had to deal with during the global pandemic, and I think we got through well, all things considered. Retention should be in your mind regardless of the state of the industry. Keeping your staff is critical to efficiency within your team. You don't want to constantly replace good people, onboarding, retraining, and hoping new staff are competent enough to do the job.

You have worked hard this past year to build a strong, efficient, cohesive team. Good work. You can certainly pat yourself on the back at this point. I want to discuss a looming problem you will still come up against. Retaining your talent.

Staff retention is a massive issue for organizations worldwide, and I want to make one very clear here. You cannot keep everyone happy. You simply can't.

The initiatives you put forward from this book are great, and when you look at the face of it, you are providing an amazing workplace for your team members. They are being trained through your development planning. They are provided with fulfilling and meaningful tasks through the responsibilities and accountabilities in the JDs. They are respected by the business for following communications and change controls. You communicate regularly with them. They even are being paid according to the industry guidelines that you researched. Still, sometimes it's just not enough.

Get this clear. It is never going to be enough for some people. They will see the greener grass somewhere else, want to climb some corporate ladder, or maybe just want a change in scenery. You need to acknowledge, accept, and prepare for that.

Retention is incredibly important, and I am sure you already know every person that walks out the door is a loss of knowledge and takes your team a step back. Replacing that person is costly, and it takes time and effort to slot in a new team member. I suppose you want a magic way of retaining staff, but I cannot provide it. I will say the following are important aspects of keeping your staff on board.

Pay Well – An obvious one and perhaps one that is outside your control. At least understand whether your company provides salaries close to where they should be. I put this as number one because it is the main factor for retention, and you need to understand it. You need to take out the salary report you created earlier and know whether your company pays well. Suppose your staff is at the standard or below. A tough conversation with team members is coming, so prepare for it by knowing this when preparing your answer.

Communicate – Talk to your team often and listen to them. Really listen to what they have to say. Too often, we forget ourselves and just babble on during meetings without hearing what our members have to say. Take it from me, listen to the gripes, try to help where possible, and always acknowledge a team member's problem as real.

Develop – Use the development planning. Follow through with it. Make your team accountable for completing it each year. They will respect and appreciate the effort and encouragement to become a better-skilled employee.

Praise – Take the time to lavish your team members publicly on their work. If they finish a project on time or earlier, let everyone know. They get certified, let everyone know. They complete a difficult or risky piece of work successfully, let everyone know. Remember, sometimes our work only matters to those in the office. Partners and family may not understand what we do or how difficult it can be, so being praised in front of colleagues is the only way to receive adequate acknowledgment for our hard work. I have found that lavishing your team with praise is fun and creates a bond.

Be firm – Flaky or unsure leadership is a quick way to lose your team. The main reason you are in this role is to make decisions. So, make them and stand by them. Unsure managers who can't decide are rarely seen as effective. Be strong and confident. But also be prepared to make the wrong decision. Now this may be confusing, but you must be able to admit when you are wrong. Be strong and confident here too. Tell people you were wrong, and you are going to fix it. Tell people you made an error and will change it. This is what builds respect and keeps the team moving forward. And lastly - When you push an initiative, see it through to the bitter end. If you make a promise, stick to it. If you decide, stand by it. People want a strong leader and will follow you through anything if they have that confidence in you.

The art of keeping staff isn't a magic spell or anything too complicated. You don't need to be super charming or attractive. I know I'm not. Through experience, I find these five principles provide the best opportunities to keep your team engaged. But you need to put in the effort. These principles are the key to a long-lasting relationship and retaining good talent. I live by them, but they still may not be enough. As I said before, people will leave for a multitude of reasons, so never take it personally and, of course, prepare yourself for it by ensuring all skills and knowledge are overlapping and covered.

Dealing with the Liars

In a previous chapter, I mentioned that you may get lied to by your staff. This is common in the world of IT. Many techs seem to think they need to know everything in this industry. That, or they think that they must know everything to be deemed competent. Sometimes your team may gravitate towards the old "baffle them with bullshit" mentality. This way of operating is a real problem, and I have had to spend a great deal of time stamping it out. Having members of your team acting like this will only hurt your department.

We know that users can be ignorant of technology, but knowledgeable people work everywhere in their chosen fields. They can smell the lies just as much as you, and when your team is caught out, it will ruin the relationship you want to improve.

Let's deal with this problem.

Remember when you started? I asked you to play dumb where you could. Just enough to get your team to open up. This is a two-pronged approach to your team. One, it will get the passion to come out (hopefully), and two, you may identify a liar.

The liar hates being pulled up on their lies, and I have found pulling them up on the spot only works against you. While you may understand the technology in the discussion, the conversation can quickly be swayed into an argument over unrelated topics where you don't have the same levels of knowledge. This is called deflecting, and you may find yourself at an impasse trying to figure out how you got here when the main topic was your focus.

So, my advice is don't pull them up at all. Update your skillset matrix to reflect this resource, in fact, does not hold this skillset and understand the person is, in fact, a liar. In your mind, you will start to question what they say, and make sure to do your research before you conduct formal design chats with them. They can quickly ruin your designs, so it's imperative that you do your due diligence in their responses.

Now you might say, why would I let the liar slide like this? These people need to know the error of their ways, so let me explain why.

I have known guys who will do the "I'll make up some stuff" answers. Too many to count, and I have dealt with it in many ways. Pulling them up there and then, maybe even in front of their peers, only gets their back up and pushes you into arguments you may or may not win. Your job isn't to argue with your staff. You are a leader, and as such, you drive the boat.

Instead, take the high road, always. Make a note of the lies, do your research, and plan your design based on good solid facts and best practices.

Make sure, and this is incredibly important, to bring back your research findings and discuss them with the liar in question. He or she will note that you have taken the time to find the best possible solution that goes against their nonsense and didn't embarrass them in the process.

Eventually, the liar will realize that lies and nonsense have no place under your command, and you won't accept it. You have also given the liar a chance to work out that they don't NEED to lie anymore. You are not a judgemental manager that needs a correct answer there and then. There is nothing wrong with not knowing the answer on the spot. You must show that not knowing something is perfectly OK, and a good engineer or tech must own up to this immediately. They will take away this new challenge and find a correct solution to the problem.

Eventually, this will filter down to the users, and your staff will provide facts and correct responses rather than lie to the users. Your department will be regarded in higher esteem, reflecting your work.

This is an incredibly frustrating aspect of being a manager, and it can be difficult to deal with. Just remember you must take the high road. Don't put yourself on the level of someone trying to look smarter than you. They can be smarter if they want, but, in the end, you need to get the best outcome for your project or service delivery.

The Toxic Culture

You may have found yourself in a workplace situation, and coming to work is either incredibly stressful or just plain depressing. You cannot be bothered to get out of bed because you must face a nightmare for 8 – 9 hours. I get it. I have been there before. The culture alone made me want to quit, give up, and look elsewhere. In fact, this type of culture made me want to rethink working in this industry. There are many clever people in this industry who are smart and know it, so they treat others like garbage. These people feel their big brains make them worthy of shaming and blaming others. Imagine what it would be like to work with the Sheldon Cooper from Big Bang Theory type. In the show, Sheldon is surrounded by friends who understand him. In the workplace, it's the opposite. These are not considered friends, and Sheldon would turn a workplace toxic.

This is one very real scenario for a toxic workplace. A single or small group of people make it incredibly difficult for those around them to grow. They shun working with others, ridicule, and even undermine their manager. You.

A workplace like this is probably the worst of situations, as others in your team will shut down, become frustrated and leave.

You may notice team members taking a lot of leave, sick days, missing deadlines, or not getting involved. Very poor behaviors in a workplace, and they may not be at fault. Attacking them for these behaviors will only hurt your relationship, so identify if you have toxic team members quickly.

Once you discover you have a toxic workplace, what do you do?

Here is my take on a toxic workplace.

Firstly, take a step back and watch what causes the toxicity. It may be several aspects; if you are lucky, it is just one. I generally find the following can cause a toxic workplace.

- Workplace whinging or gossiping
- Narcissistic behavior
- Lack of meaning

Take note of which of these is relevant to your area. It may be a few, and it may be difficult to pinpoint but make a note of it. Some may also be linked to others or caused by others. No problem. The first step is to identify that there is a problem.

Here's how to deal with each one.

Whinging or gossiping

Just about every place I have worked spends time gossiping and griping about other people and business processes. Seems to be commonplace when people get bored. Putting people together in any location and the instinct is to gripe about something.

I don't think it can be controlled. It is going to happen, and lesser minds will get involved. I make this statement because you must ensure you do not get involved. You need to stand above this griping and keep a cool head when surrounded by complaining.

Your first step is to ensure you don't get involved, but the next step is to try to control the whinging. To do this, start by finding the core of the whinging and take them aside. I have found there is usually one or two people that have the respect of the rest, and when they start, the rest follow. Make a note of these people and take them aside. They need to be told to gripe up and never down.

It's as clear as that. Gripe up and not down.

Even though they are seen as peers in the company's hierarchy, they are still looked up to by their peers.

They must limit their whinging or griping to you, their manager. Make this a KPI if you must. Challenge them to limit the griping to you and only you. It may require many small "chats" where they will offload their problems on you, but it will slowly bring the team to a respectable state. Soon you will find these guys actively stopping themselves from the echo chamber of whinging and just getting on with it.

The gossiping, though, just happens. You just need to ensure you don't get involved. Gossiping can lead to serious problems, especially if the men begin making comments about women or vice versa. Do not get involved. Keep above this and focus on the job. There should be plenty of work, so keep your team busy with the tasks previously determined in the book.

Narcissistic behavior

Narcissists are a pet peeve of mine. I can't stand these people and never really understood the need to act this way.

But they are out there. These guys feel it's their way or the highway and do not buy into other ideas. Zero empathy, zero compassion, zero understanding, and, I think, bad memory. We all had to learn at some stage and at different speeds, something the typical narcissist cannot seem to remember.

I have worked in a few places with these guys. Generally, they have been around for a long time and know the systems back to front with a lot of blood, sweat, and tears put into their setup. I get it. How dare you question their work? Any ideas that fresh eyes bring on have no benefits to them and only seek to undermine the many years of work. Trying to argue any point with these guys can soon make people feel lacking in skills, knowledge, or competence. As a manager, you can feel undermined at every turn. "We don't do it that way here" is a common adage.

Arguments can quickly arise, a need to one-up each other can occur, and belittling others' mistakes will take place. Very toxic indeed.

So, what do we do about these narcissists? How do we get from a place where everything needs to be approved or amended to suit these horrible people? The answer is process and governance.

I don't deal with these people directly and begin arguments. You will quickly lose in a battle for knowledge. In some cases, these guys have years of experience in this environment. You won't win that fight. Your answer is the compliance documentation you have created in previous chapters. Your systems must comply with these and be secured in a certain way. Rather than arguing with individuals about the best way, use your compliance requirements to provide a foundation for building best practices.

So, you have completed your audit of the systems and see gaps, or you have engineers arguing over a change or design piece. You can stop these arguments by relying on your compliance documents. Use these as your guiding light.

I'll give you an example of my most challenging experience.

I was hired to provide direction for a small ICT team where the lead engineer had built up the infrastructure, on his own, from the ground up. He followed what he thought was best practice and made decisions based on some Google searches and a large intellect. He was an incredibly smart individual but soon found himself in over his head. The organization decided to hire more staff to help him manage the systems. Soon he was surrounded by very clever systems and network engineers, yet, he still fought tooth and nail to configure systems in his own way.

A narcissist through and through, he managed to win every argument with "we do things this way here" or a high level of knowledge of the systems being managed.

The culture was toxic when I turned up. Very toxic. I was overwhelmed with the nasty way change was dealt with and the refusal to improve. "We don't do that here" was a daily response.

So, my savior ended up being a compliance document created for the industry we worked in. It provided statements and controls on how systems should be secured and deployed. It was black-and-white documentation handed down and provided me with a way to audit the systems against it.

I ran my audit and found the setup had multiple gaps in security and regulation. I found that, in fact, the best practices have not been followed and found that most industry standards were missing.

From there, it was a simple case of working with the team to close those gaps and the path of least resistance to becoming compliant.

In this scenario, the narcissist refused to assist and made the same comments repeatedly. "We don't do that here," "This won't work for us," "That's not possible," etc.

We pushed on with our findings and began designing the perfect infrastructure. With minimal input from our friend, we started forming a decent design for the future of our systems.

Not long after this, the narcissist seemed to realize his shortcomings and left the organization. It was unfortunate. He was a talented individual, just not a team player.

You can see that processes, best practice design, and compliance saved us from a lot of heartaches. When dealing with the argumentative types that only see things their way, you can always rely on solid documentation to support you.

Lack of meaning

You may find yourself in a culture where the team feels like a cog in the machine, or individuals who are bored with their repetitive work may feel a little like robots. These are the guys who want to be informed of what is happening, even if it is not a task directly involving them.

When you carry on business without their input, they quickly become negative and start making waves about "no one listens to me," etc.

The need to feel valued is incredibly important, and I feel it is our job as managers to ensure our team feels valued. You don't want superstars looking around for a new job because you cannot make them feel part of the business.

While having members feeling this way may not turn the workplace culture toxic, this environment can quickly lead to negativity in your team. Having people feel like they are being used as a robot will turn them off quickly, so it's important to work through it. And this is an easy fix. Your spot meetings, which we mentioned previously, are incredibly important to ensuring your team is involved. Give each team member a chance to speak their mind and get involved with different projects.

Take your KPI process seriously. Make sure you discuss your plans, listen to their needs, and always act on your promises.

Always look for opportunities to mentor your team, teach them what you know, or have the junior members work on projects with the senior members. Look for ways to challenge your team at every turn.

It's easy to become comfortable with your team running repetitive tasks and keeping the business running, but you can't stagnate with the challenging stuff.

Project Management

I know you may think your job has nothing to do with managing projects. But you are wrong. Many IT managers are hired to manage IT but quickly find that managing IT includes managing IT projects. Projects that are driven by the business with IT requirements left off the list.

Let me give you an example. A new project is raised internally that you are unaware of yet. The project requires some new business processes. Part of the new business process is entering data into an operational system. Here is the problem. Sometimes Project managers will forget that this operational system runs on IT. What should have been included as part of the project deliverables is left out or incorrectly scoped, and suddenly the project is at risk of failing.

Also, many IT managers are given a piece of work that is seen as BAU but is, in fact, a full-blown project. Sometimes Project Managers get it wrong or lack the competence to engage your team initially. It can become your problem to manage a part of the project yourself.

Whether some work is seen as a small change or a major infrastructure upgrade, you need to get your head around project management.

I am going to try and simplify the whole thing here. I have watched many hours of project management videos and sat a few courses on the best way to achieve project success, but here in a brief nutshell, I will explain what you need to work out for your areas to achieve good results.

You can spend hours creating documentation. More time designing a small bureaucracy around a project. A lot of project managers do. They get paid by the hour; I am pretty sure. I will simplify it so you can get your head in the game and know the basics of running a project.

At this point in your career, I simply wanted to explain this project management stuff, so I will break it down into six stages. It can be a lot more complex than this, and you can add a lot more training to your reading plan, but for the purposes of this book, ill make it a simple start for you. So, I want you to focus on the six stages and get an understanding of them if you are expected to manage a project as part of your new job.

Target StateFirstly, there is the target state or the goal. Get this part as accurate as you can. Find out what you will have at the end of a project and who will use it. I cannot stress this enough. Get the main goal of your project written down and accepted by those who will use it. I'll give you an example. A middle manager at your company thinks it would be a great idea to set up a system where accounting will clock in and out of a time management system to allocate hours against certain tasks. He comes to you and says you need to implement this system because he and his team think it will help Accounting. Great idea in his mind, right?

Well, let's take this project and run with it. One of your team sets up a VM, allocates storage, installs the app, deploys the client, pushes the software out to the accounting team computers, sets up backups on the new service, hooks up monitoring for it, and your staff creates documentation to support it. All this is money, time, and effort, and you successfully get this app to work. Great!

After monitoring the VM, you realize no one is using it. A few entries when it first fired up and then nothing after a few days. You quickly find out that the new service isn't even needed. Accounts can't see any benefit in clocking their time. In fact, they don't work the way the middle manager who initiated everything thinks. What a waste of time.

All of this could have been resolved if someone had just taken the time to discuss this with Accounts. I know you may think it's not your problem, but now you have a system in your environment that wastes resources and money. So, is it your problem? I think it is. Many IT managers may ignore what happened and move on, stating this is outside of their remit. I don't agree with this way of thinking. I think you should care about what is deployed, its usage, and its benefit. That middle manager's great idea was a waste of your time. The time could have been spent on a project benefiting the organization or resolving real environmental issues.

So, as I said, find out what the end state will be and talk to the users who will use it. Always. I don't care how high up you get. Take the time to go and speak with the end user and see if they will benefit from someone's great idea.

Scope Definition

Now that a project has a clearly defined target state and user buy-in, it's time to discuss the scope. Find out what exactly is expected. I might be confusing you between scope and target state. That's fine. The goal of the project or end state should be a one-liner like "Users will be able to clock their time" The scope is the details of that. So, items like, Users will authenticate via SSO, users will have access from any device, the server will be cloud-based, etc.

The scope is where your experience comes in, and you can sit down with the project facilitator and go over the project's particulars. This meeting or set of meetings is called data gathering and needs to be written down or summarised. Send that summary to the people who attend, and should no one question it, this is what you work towards. If there is any response directly to you along the lines of "Actually, I want this or that added," you need to let everyone know or set up another meeting so you are all on the same page. This scope you are writing up is critical to the project's success. In this industry, and probably most industries, scope creep is a real problem. I am sure you have seen it before where you work towards providing a system, and random users will come in and want some part of it changed, or some feature added that you didn't agree to in the first place. This is called scope creep, and you should avoid this. Tell these people that you are working towards a scope, and if you want changes to the scope, it will blow out timelines, costs, and maybe even the target state already agreed on…

The scope allows you to stay on target. The scope allows you to say no, and most importantly, the scope allows you to design a solution. That's where we go next.

Design

Now you have your target state, buy-in, and scope for this project. Next, you want to design the solution. This is the cool part. Sit down with your engineers or admins and explain the target and scope and then begin designing how it will look. Whiteboard your ideas and come up with the right solution. Talk to vendors if you need to, as there may be new ways to provide the solution, but at the end of the design phase, you should have an agreed way to provide the IT for the solution.

Next, you mark up the cost involved, and this is where you get your design may have to go back and forth. If your gold standard design is knocked back because it's too expensive, go back to the design team and work out a new solution based on the acceptable cost to the business. This process can get old quickly and frustrates many people, but you must stick with it and keep your team engaged the whole way. They need buy-in on the design; if the scope is met at the end of the day, you are doing your job effectively.

Once you agree on the right design for the right price, the next is to allocate resources and time to the project. This is called a programme, and here's how I do it.

Programme

I create a spreadsheet. I take the project and break it into individual tasks in the first column. These tasks are what a single person needs to do. In the second column, I write how long it takes to complete the task. To do this, I talk to the resource involved or their boss and ask them. "Hey Jess, how long does it take you to create a DNS entry" Jess will respond depending on the type of person she is, and if she underexaggerates, double the time. If she exaggerates, talk it through with her. Find a happy amount of hours/days for the work to be completed making sure to remember BAU workloads.

Once you have the timings under each task, you can see how long the project will take holistically.

Some of the tasks will be deemed critical and cannot be moved or adjusted in any way. The project cannot move on until they are done. These are called tasks on the critical path or key milestones. Highlight these.

In a third column, enter milestone targets, deadlines, and the month and year they will be achieved.

In the fourth column, place the budget to be spent at that point, and the fifth column for the vendor or supplier to spend with.

Below is an example. A little exaggeration, but you should get the idea. When it comes your time, feel free to add more detail where you see fit, additional columns if they make sense, but the key is for you to walk away understanding how long this project will take and at what points you will need money to spend on it.

Task	Labour (Hours)	Resource	Deadline	Budget	Vendor
Configure networking	1	Jess	Mar 2022	NA	
Assign DNS	1	Jess	Mar 2022	NA	
Add IPAM entries	2	Jess	Mar 2022	NA	
Purchase SSL	1	Tek	Apr 2022	$200	Easy SSL
Configure Firewall rules	2	Hamad	Apr 2022	NA	

Contingency

A hot tip here is once you list all your tasks and add the hours give yourself 20% on top of the labor. If you are wondering why we added the 20%, this is called contingency, which is key to projects. You need to manage expectations. When you give people an expectation that a project will be released quickly, of course, they will be happy you said as much. But watch their respect for you dwindle when you fail to hit that promise. On the other hand, when you tell stakeholders about a realistic timeline with 20% on top, yet the project finishes early, you are seen as a hero. Just remember you don't want to lie to anyone. You want to be honest, but at the same time, you want to give your team enough time to make mistakes, learn and test their work.

Take team members through the plan and ensure they are happy with the allocations and expected timings.

Shuffle things around if you need to but make sure you have a solid plan that can be shared with all stakeholders.

Now you have your scope, you have written up a pretty good timeline, and you have a budget to complete the work, let's have a kick-off.

Kick Off

Make the kick-off something special. Let people know the project has been kicked off and when to expect it to be finished. Keep on top of the timeline and make sure the right people have what they need.

I remember a place where anyone working on a critical task would put a red balloon on their desk to show everyone else they are working on a project and to be left alone. Other places put a little sign on the back of their chair explaining they need to be left alone. Let your imagination run free, but as I said, these people made promises about how long it will take them to complete these tasks, and now, they need to deliver. You have that 20% across the project to deal with hiccups, but your team doesn't need to know that; they have a deadline to meet.

When tasks are completed, congratulate those involved and inform the stakeholders that the next phase is kicking off.

Keep going until the project is completed, and you can continue ticking off the original scope. Remember when I mentioned scope creep? As the project is in motion, ideas will fly around with people trying to scope creep but stay the course. Get the project completed as promised and on time. Sometimes you cannot avoid scope creep or items outside your control. You can deal with these by requesting a variation to your scope in the form of a project extension and/or a variation payment. You must inform all stakeholders that this scope creep impacts your delivery and timelines. It's not free.

Completion

Sometimes it is unclear when a project is completed, but a good project manager can identify its completion and what that means. You will need to draw the line at some point and say the project is complete, and we are now into operational support. This is critical as you don't want the project to drag on, and you don't want to be seen as failing. So, grab the scope you did in step two and start ticking off the points in there as done. If you come to a point where you cannot tick off an item, work out whether it's due to something incomplete or something that cannot be done.

If incomplete, get it done. If it cannot be done, determine why and ensure this is communicated. Your scope makes the project deemed finished, so go back to it and check you completed everything as discussed those weeks, months, or even years ago.

When the project is completed, whether on time or not, take everyone who was part of it out for a celebration. Even if it's only a Pizza lunch in the conference room, do it so those involved feel they have finished the project.

Conflict Resolution

This is a difficult topic. Suppose you are an introvert like me or the type of person who avoids conflict. In that case, you probably become apprehensive when an actual conflict occurs, be it between two other staff members or, even worse, between you and another.

While I can never expect to teach you how to be fearless, I can give you a few pointers.

The key to understanding and resolving conflict is stepping into other people's shoes. Take the chance to look through other people's eyes for a moment and try to see things as they do. Unpack what is really going on and why things are escalating. Figure this out first before judging or determining your next course of action, and if you do not know, ask. Ask in a way that does not escalate things further, and ensure you are open to both sides even if a side opposes your opinion.

I find there are usually very few reasons people become abrasive, and ill list them out.

- They are a narcissist.
- They are not being listened to.
- They are just having a bad day.
- They look for conflict and enjoy antagonizing.

Not much to it. These points are at the core of most conflicts, and it's important to understand which one is causing your current problem.

Now we understand a little about why people get into conflicts, let's break down how to deal with the actual conflict itself.

- Clearly define the reason for the conflict.
- Get everyone to take a step back.
- Listen to all the options. Allow all parties to present their case without interruption.
- Seek alternatives to bring to the table.
- Come to an agreement, or you make the decision.

Let's unpack these by using real-world scenarios.

Let's say the conflict you are dealing with involves a couple of engineers on your team. Sean and Pete

Sean and Pete are arguing over the best way to deploy a new InfoSec policy, which is getting heated.

Sean wants a flexible policy allowing users freedom on their workstations, and Pete believes in a more locked-down approach.

Now you have your opinion on this topic, I am sure. We all do, and I am glad you are picking a side already because it shows how difficult this can be.

Let's salt this problem with the characters. Sean is one of your senior employees and has been at the company for over a decade. Sean retains knowledge like a sponge but is a narcissist. His way or the highway. You came on as his boss, and he has never really supported you.

Pete is very junior. Pete works 10–12-hour days to try and catch up with his peers and looks up to everyone, including you. Normally he brings up these issues and defers to others, but this time he is adamant and standing his ground.

Now you may be starting to sway your opinion on this topic. Sean must know, right?

Well, that's not the point of this chapter. We are here to resolve the conflict, and it's not about who is right or wrong.

You need Sean to see value in Pete's opinion, and you need Pete to understand where Sean is coming from.

So, let's deal with step one here – Define the reason.

We know it's about infosec policies, but that's about it. In fact, it may not be about that at all. I have seen people arguing over random issues before, and when trying to unpack them, we quickly realize the solution is focused elsewhere. That may very well be the case here.

"What's going on" is always a good way to start.

Now at this point, the argument may continue with ego involved and Sean talking over Pete or vice versa.

"I can't follow when you talk over each other. Sean, can you start at the beginning"

Now this is where Sean gives you his side of the story. Remember, it's from his perspective and does not contain the whole truth, just his side of the truth, and to make his argument the most valid.

He says Pete is trying to ruin the user experience by restrictive policies that won't let them do their jobs. The policy that Pete wants to deploy will make everyone's life a living hell. The only way to deal with this is to deploy it using Sean's solution, and he finishes up with a whole bunch of technical solutions and acronyms. Certainly, sounds like he has the answer. Always does.

At this point, Pete is smoldering. Make sure you give him a chance too. Never ignore a party involved in the conflict, or you will be in danger of favoritism or worse.

"Your turn Pete."

At this point, Pete may try to argue against Sean's points. He may say that what Sean has said is wrong and focus on Sean's comments. Listen to the points he makes and stop Sean from interrupting. The idea here is to find out the reason for this conflict, not allow it to continue.

Pete is saying that Sean is adding risk through his solution and being overly lax in his approach to cyber security.

Time to take a step back. This is step two and incredibly important. Both guys have valid points, but what are they really arguing about?

"Where did this request come from?"

Either one of the guys can answer now. Where it gets interesting is if their answers are different. That means you need to take a further step back or listen and listen to both sides again, but in this scenario, they both agree the firm's engineering department requires a new piece of software and logged a ticket to have it installed.

Pete picked up the ticket and asked Sean to deploy a new package, as the software requires local admin privileges. The packaging will protect the SOE and ensure vulnerabilities will be mitigated.

Sean saw the ticket come to him and is trying to refuse the work, claiming it will restrict the engineering department, which is used to flexible IT policies and has admin privileges when needed.

This is a touchy scenario that all IT managers must deal with. The fine balance between usability and security.

Now you may or may not have written up your security policies for the company or have a security specialist within the company, and that is why this is a point of contention here. Still, both members have valid points.

The point here was that both were allowed to clearly define the reason for the conflict, and you took a step back to unpack why it had gotten to this point. You are getting there.

The next step is to table solutions. You may have them already in that we open up for the engineering department, or we package the software, but there may be other options. This is where we look for empathy from each member.

"Are we able to deploy this any other way that doesn't compromise us, Sean?" try to get him to see Pete's concern.

At this point, if it was anyone bar Sean, we may get an open dialogue where we look at alternatives to just opening up the system, but as I said, he is a narcissist and only sees things his way. So, as close-minded people do, the response will be the same, backed up by some technical jargon to try to back up the original solution. Don't let it phase you, and stay the course. There are always multiple ways to solve a problem, so find them.

"Pete, do we have any alternatives to packaging the solution?"

Pete being fresh, will probably come up trumps, but he may surprise you. He will probably request time to research or go back to the software vendor for a fix and get back to you.

Either way, we need everyone involved to see all the solutions. Sean gave you one. Pete will find more than one.

After a day or two, Pete re-opens the discussion and says the software vendor has a client/server model which will work for us and not compromise our security. After discussing it with Sean and yourself, it is agreed on the best way forward, and the last step is complete.

Remember that the key takeaway is that your job is to get all parties to understand each other and make a mutual decision. Not all conflicts have a happy ending for all parties, and in the next scenario, we will do just that.

On a side note, Pete comes to see you and makes you aware that the packaging of the software is not a great impost for the company, and it's something Sean should be doing as part of his job. You know and want to acknowledge it for Pete, but this will further sow discord. Sean's narcissism is having an extremely negative impact on the office. Ensure this event is logged, as it will be a growing list of issues you need Sean to address when it comes time for a performance review.

The squeaky wheel

You have probably heard the term; the squeaky wheel gets the grease. Or to help those that shout the loudest. You have probably dealt with this issue repeatedly, and it becomes routine to deal with the issues that impact the noisiest complaints. This contributes to delays in what is important or planned work. It's a bad habit to get into, and allowing it to occur should be avoided.

I don't need to give you examples of the negative impacts. I am sure you get it and want to know how to stop it. So here goes.

The first step is to ensure you have the right processes in place for providing support. The ITIL initiatives you set up in previous chapters should have put you on the right path for these processes. You need to have a good ticketing system, and you need to have a good prioritization process.

If you have these in place and are working through everything, your next step is ensuring the system works. Review several incidents and see if anything is tying up the process. Ensure that information flows well between your team and the person who has raised an incident.

Try to keep communication lines open all the time so that the end user is getting notified of steps taken to assist them. This is the key to successfully supporting your users.

Now with these systems in place, it's just a matter of removing the expectations that people will jump at your command.

It is two-fold. You need to ensure that issues are raised properly, and those same issues are managed properly. You can't fix the squeaky wheel problem unless these two are handled appropriately.

So, you have your systems and are managing the incidents correctly with plenty of communication back to the users. Therefore, you should have confidence in your system. With that confidence, you can return to those who shout the loudest and challenge them to use the system. Log a ticket, you say, you will be surprised. A lot has changed, and I will ensure you get what you need.

I would advise against telling noisy users that they are holding up important work by not following a process. They may resent you for putting a negative spin on their request. Just stick to the system you believe in and watch it work its way into the business. If there is any part of the system, you do not have confidence in. Fix that first and foremost. Ensure your team is squared away with the process and get that confidence.

Pretty soon, over time, you will find there are no more noisy users shouting over each other to get their "urgent" work done first. They will get the same level of confidence as you have in the process you have set up.

I know, I know, there are always annoying users or clients that find themselves more important. Well, you are right. And you need to put yourself between your team and those users and request professionalism from them. Protect your team by answering those requests and always sticking by your process.

Presentations

I want to talk a little about how you present ideas and directions to your organization's peers and upper management. Whether you like it or not, you still need to sell your ideas as a manager. You need to present ideas to non-technical people, and you need to present your ideas to people who may just see your department as a nuisance or a money drain. I have had many instances where I am trying to get money for a solution the organization direly needs, only to be disconcerted by the fact that my audience doesn't understand. That, or their eyes glaze over from the boring topic I am presenting.

So, let's say you have looked around your new organization and found that the network equipment is ancient, slowing down key services and probably going to fall over. You can go up to the CFO, CEO, or whoever and start talking about packet loss or jitter, but they don't care. They don't. Nine times out of ten, these guys just want to know how much this will cost and the benefit of buying it.

You may be lucky and have a technical lead that you are explaining your need, but this is rare. So, I will provide you with the key points to ensure you get what the organization needs.

So, going back to the example, network gear. It is due to fail, and you must ask for serious spending to get the right upgrades.

Firstly, determine how the organization buys for its large projects and who approves these spends. Ask around to see who got money for what and how they did it. You may already know this, but you want to be sure who your target audience will be when you present your proposal. When the time comes, do you have to impress the CEO, CFO, a board of directors, or your line manager?

Secondly, don't mention anything about the issue to them yet. Going and bleating about what you will do without much forethought can quickly derail your proposal. In this example of network gear purchasing, say you go to your line manager and mention, "Oh hey, I'm going to be asking for some network gear," and when they ask for details about it, you fumble through your answer. In the back of their minds, they will quickly feel like you will waste their time. I have seen it time and time again. Techs want something, they carry on about it without answering the benefits question, and managers will roll their eyes and think, "Another IT toy! Won't be wasting my money on it…"

Don't mention the issue until you have researched and are prepared to answer every question accurately. You need to think strategically here and, as I said, put your sales hat on.

Thirdly, research, research, research! Take the time to look at the problem from all angles. Do I really want enterprise gear because my engineer says it's the best? Maybe a cheaper option will do the job. Is the cloud the best place to add my workloads? Does this contract add benefits? You don't know yet, and meeting with vendors is a big part of your job, so do it. Get out there and tell them you are looking at a major upgrade or project. They will happily sit with you and your technical expert and review your options. I have had engineers 100% against vendors like Cisco. "Overpriced" or "Useless" are terms you can hear, but once you put these guys in front of the vendor, they usually relent that their solution is a good one. Sit in front of multiple vendors and do the research needed to ensure you get a few options.

Put your option and everything you have learned into a professional document. Call it a Business Case or Proposal or whatever term your organization uses to request funding. At the same time, you should get your PowerPoint skills brushed off and start creating a presentation relevant to the audience. Remember, the first step in this process was identifying who could approve the spending. Your presentation needs to target them. Are they technical people? Well great. Write your presentation with the technology benefits and how you will improve performance. Are they non-technical people who want to know how value is achieved? Good. Target your presentation to make them understand the productivity benefits. The key to winning these people over is to have everything ready and show you are prepared.

Lastly, book a time with the people that need to know and present your PowerPoint. Hand out the business case you created earlier, and be prepared to answer all questions relevant to your proposal.

Public speaking takes some getting used to, and each organization does it differently, but you should jump at every chance to do these presentations. Get good at it, and you will make yourself incredibly valuable in this industry.

Public speaking can be daunting, but if you look at all the effective and respected managers, they can get up and put their thoughts forward effectively and confidently. That needs to be you, so practice in the mirror, practice with your wife, husband, partner, kids, or even your dog. Just nail what you want to say, book the meeting, and win that approval for funding.

Business cases

An absolute must for any manager is to be able to articulate your plans and moves professionally. Talk is cheap, and having great ideas and how you want to deliver them is only part of the picture. You need to be able to present these in a document so that anyone with a little business knowledge can pick up and understand.

There are many factors when putting together a business case, and I am sure several books and blogs detail every aspect of it.

I am here to give you a few pointers; you must take it seriously. It is an integral part of your role and, in many cases, a representation of your skills.

Let me explain.

Until now, people's expectations of you may have been to provide direction on IT to your team or ensure operations are running smoothly and steadily. A lot of the guidelines in this book will help you achieve that, no doubt, but you will always cross a bridge where major works or changes to the business require understanding and funding to be obtained. Until now, you may have presented your ideas to your manager and relied on them to make a case for you. It may have been as simple as bringing up what is needed via a meeting or a chat by the water cooler. Your boss took that info away and delivered it up the chain, funding was provided, and you delivered a project that improved the business.

Now I am by no means going to assume your boss has taken all the credit here, nor am I going to assume you are happy with this arrangement. I want you to understand that the process of delivering the idea should be yours to take end to end. The idea should be yours to present, yours to own, and the credit should go to you. After all, you only need to create a business plan that articulates the idea professionally and covers all bases. You will learn a lot from the process, and who knows, maybe as you go through it, you will find out your idea isn't so great after all. Perhaps the ROI simply doesn't add up for the organization.

Let's get started.

A business case is all about presenting your idea; like all communications, it should be written in a way that the audience will understand. You should put yourself in the reader's shoes and think about what they will want to know and what they will need to understand. Most importantly, how do I make my idea appealing to them? Remember, your idea is awesome for you and your team, and you get it, but the people with the funds may not have heard about it before. You need to try and take your headspace to the place of non-technical people who count numbers all day and worry about budgets and take them on a journey by creating and presenting a document known as a business case.

To start your document, don't worry too much about formatting. If your organization has a template, forget it for now. Start with your blank Word document and fill out the following headings to get you started.

- Executive summary.
- Document purpose.
- Current state.
- Target state.
- Benefits.
- Cost.
- Market Scan.
- Estimated return on investment.
- Risks.

These always put me in a good headspace, as now I know it's all about filling each section out with the details of your idea.

Executive Summary

Start with the executive summary. This needs to be a very high-level explanation of your idea. Think of it as how you would pitch it to someone on a walk down to the local café or pub. You don't have much time and need to explain it in a few words that cover most bases. This should not be a full deep-level explanation, just a summary, so keep it to a paragraph or two. You will return to this section many times as you do your business case, as it will be adjusted to suit the detail below. Start here and nail the idea as concisely as you can.

Document Purpose

This section needs to cover that this is a business case to explain your idea and who it is aimed at. A paragraph, no more. It is not a further explanation of the exec summary; it is a purpose. Meaning it lets the reader know who it is meant for and why the document (not the idea) is needed. This may be confusing, so feel free to leave this until later, as it will make sense. At the very least, put a list of the people in your target audience.

Current State

Now you need to make the reader understand the current state of things. Keep things factual, accurate, and, above all, understandable. Keep yourself in the reader's shoes and ensure you are writing to them at their level of understanding. Start them on the journey. Write this section as an educational piece of literature. If you are writing to the CFO and CEO, for instance, they will want a high-level understanding of the risk to the business. What have you found that requires their attention? What part of the business does it affect? That sort of thing, in a way it is presenting the risk, but you need to write it so it's understandable. Vomiting out technical jargon is not understandable. Whinging about previous mistakes is not going to help either. Remember, factual and accurate information is what you need to write in this section and at a level your reader will understand. This section may take a while to write up. It may need diagrams, and it may even need pictures. You are going to need to make it clear. Take your time and leave nothing out. Try to cover everything your that idea will replace, resolve or improve.

Target State

Now the fun part. This is where you define your idea in its entirety. Like the previous section, you will need to remember your target audience. Stay at their level. Write your idea to get them excited about the changes that are coming. Write it to make them want to deal with it immediately. Sell it. Again, pictures, diagrams, and glossy terms that keep people interested are what you are after here. This should be the easiest part of the document for you as this is what you have had in your mind before starting this document.

Benefits

Where the previous section was all about the target state and how it will look, it is about the actual benefit to the organization. Again, think about your audience. Is this the CFO who wants to understand how much money is coming in? Is this the COO who wants to understand how efficiencies are provided? Is it all of the above? Keep them in mind when writing this section, and answer them all. Try to think up a list of their questions and answer them upfront by nailing the benefits to them in this section. Imagine them nodding away as they read through and saying, "Yep, that is a problem we needed to be solved," or "How did we not resolve this earlier." That is what you are trying to get across in this section.

Costs

The all-important cost section. Avoid placing a bill of materials here. Instead, focus on a few key areas your idea covers and place a cost against each. You may have options here, but be mindful that when placing options in this section, you must explain them in your target state and benefits section. You will also need to explain them further in the risks section. It depends on how you want to propose this. Some people like to have options placed in front of them and the ability to choose one, whereas others will get confused and want a simple yes or no choice. I prefer to avoid giving the opportunity for people to say no and instead provide several options to them.

Market Scan

This section shows you have put your due diligence into the projects and reviewed multiple solutions and vendors. Here you can put all the options you reviewed and why you chose the solution under the target state.

If you spoke to multiple vendors, explain who they are, what their solution would have achieved, and why you selected or rejected them.

Estimated Return on Investment

Here you must quantify the return on the investment you are asking the company to make. You may think there is no return, but there always is. Think about how productivity can be improved, how efficiency is achieved, or how power consumption is reduced. These are just some ideas, but you need to know what financial benefits you are bringing to the table.

If your request will sell more of a product to your customer base, this section will explain how that will be achieved.

Once you understand that there is a financial benefit, you need to work on how long it will take to recover the cost of this request and show your calculations.

There may not be an ROI, this is unlikely, but you can always explain that this request is to manage what is called technical debt. That is, you are refreshing aged systems that are overdue.

Risks

In this section, you must cover the risks of not implementing your request. You will need to be very clear about the potential impacts of maintaining the current state. You will need to articulate the financial implications of not addressing the risks. These are usually put into a table known as a risk register, with the risks ordered from highest to lowest.

You should score your risks as High, Medium, and Low and then provide an overall risk summary stating your thoughts on ignoring the main issue.

You should also add a section here to cover the risk associated with choosing one option over another, as mentioned in previous sections of the case.

Lastly, talk about the risks of the actual project itself. You want to show that you have thought about the risks of rolling out your idea. Show how you have thought about the project's risk and what you are doing to mitigate the risk.

Putting together a business case is no easy task if you do it right. But as your career progresses, you will get better at it and understand the purposes of running through one. It is an important and valuable process to go through, and as you write up your case, it will help to solidify your idea. When you have to raise a business case, set yourself plenty of time to do the work, as it requires a lot of research, correspondence, and writing.

Micro-segmentation

This section will discuss something I like to implement wherever I have worked. There is a lot of discussion around consolidating systems and services, but I like to do the opposite where possible. Micro-segmentation separates services as much as possible to control individual portions of it. You can look at individual services you provide and make a strategy to separate these out as much as possible. Try to think – How can I split this up so that I can minimize the impact when we need to work on it? So, if we use an example of a software system, you have a database, an application, and a front end. You should be able to host these on individual servers so that any issues with either of these services can be worked on without impacting the others. In a more general example, historically, some servers hosted multiple services on them at once. Databases, email, antivirus, and domain controllers may have been hosted on the one box or virtual machine. You should note anything like this in your systems and plan to separate these where feasible. Imagine an issue with your database that requires a reboot of that server. You had to bring down email, AV, and domain controller to complete that. You shouldn't have to accept an environment where these situations can happen. Do what you can to segment out these services and do it as soon as possible.

I'll give you a great example. In the early 2000s, Microsoft sold a product called SBS, or Small Business Server. These were selling like hotcakes because, as a small business, you could build a single server on a small amount of compute that did almost everything your business needed. The license allowed you to install a domain controller, database server, email server, and file server all in the one box. Sounds crazy now, but it was a reality for a lot of places. I was working for an MSP at the time, and there were a lot of these SBS instances that we needed to support.

The amount of time and effort we spent on these was phenomenal. I mean, it was great for the business of managed services but a nightmare for the businesses having to run their operations. You see, if there was a single server failure, the whole business went down. If you wanted to patch the email instance, the whole business went down again. If a service stopped running and a reboot was needed, well, you get the picture. Not great for maximizing uptimes.

I hated working on these SBS systems, and that hate has flowed into my psyche. I look to segment systems in all my design work and ensure that there is never any system that will make other systems vulnerable.

Reporting

Another good step in your communications plan is the creation of reports. You should gather info about key systems within your remit and measure their performance and health.

Whether your management cares or not, it is incredibly important for them to know they are getting value from the IT systems. And reports are a good way to discuss issues or systems needing TLC. I always structure my reports to provide the following info.

Key systems and their health

Under this section, it's a good idea to put pie charts or cool graphs showing these systems are running as expected, and outages are presented with reasoning. You can use solutions to generate a lot of this data. There are some great options out there, like SolarWinds or Kaseya.

Current projects and their progress

This section can be presented with a nice percentage graph and some text explaining key deliverables or milestones obtained.

Recent expenditure

Some line items here explain what was spent and what it was spent on. Add some detail as to the benefits of spending it.

Tickets opened vs. tickets closed.

Easy one if you have a good ticketing system. A couple of boxes with the number of tickets opened vs. closed. Add a comment on why there is an increase or decrease.

Major issues

Add just a paragraph per issue, explaining what your team has discovered or resolved at a high level that impacts the business. No need for war and peace. Just explain what is going on, how you are dealing with it, and/or recommendations for future changes to avoid it.

Anything outside of these is generally fluff. You can risk adding more details if you want, but remember that you don't want your audience to get bored with the information. It needs to be sharp and to the point and reflect what a great job you and your team do.

Me time

An incredibly important aspect of life is to ensure that you balance your life with your work in such a way that you can keep your sanity. Ensure you don't burn out from your work or become a stressed-out mess. Becoming a stressed-out mess or burning out needs to be avoided. Research has shown that stress is a major factor in health-related issues. I had seen it firsthand when I was first starting in this industry. I spent way too much time at work pushing myself for a company that didn't care, making deadlines I set for myself that, from a different perspective, really didn't matter. I worked late nights, answered calls whenever they came up and studied technology that I had no interest in, and if I was to tally up my hours spent on work, It would be depressing. It is important to make sure you avoid living your life like this. I get it you want to show your value and be the go-to guy for your company, but this is not the way to do it. Trust me.

At a certain point, I realized working the way was a quick path to burning out, losing my wife, or worse, having health problems. I was having health issues, and I never realized it was a direct relation to the stress I put myself under. The keynote here is the stress I put MYSELF under. It wasn't my job or any other part of life that did this. I did it to myself.

So I changed it.

I decided to see what would happen if I took a different approach. I wondered, "Will the world stop spinning if I didn't answer that call?" Will the sky fall if I don't work 14 to 16 hours? So, I tried that on to see what would happen. I quickly learned the world didn't end; the sky didn't fall, and I didn't lose my job.

Instead, what I found was that I had more time for myself and more time for my wife. I could focus more on the other parts of life that I enjoyed. I could unwind properly from work, improve my health, and become a better person. Work was not the all-consuming aspect of my life, and I was generally happier.

From the work perspective, though, I found people adjusted too. They respected my changes and were able to work around me. I was able to jump into my work eager and refreshed each day. I could focus better, empathize better, laugh, and still get everything done. There weren't any negative aspects after the adjustment happened. I won't lie, though. At first, there was an adjustment that needed to happen. People expected I was available all the time, and I was just happy to fix problems and make decisions all hours of the night. That is just what they were used to. Not answering the calls and emails after hours and leaving all my work behind on the weekend took a while to sink in. Both for me and the people I worked with, but eventually, it became the new norm. I was no longer seen as the guy 24/7 guy.

It was glorious after that. My time was finally mine. I was able to take back my life.

I want you to remember that for yourself. It's very easy to become the other guy, The one that works all hours of the night and thinks the value is in the amount of time they are at work. It's easy to fall into a routine where work becomes all-consuming. Some companies just have that kind of culture. But I want you to know it's not necessary, really isn't. I have worked in multiple organizations and found that none have considered my work-life balance a problem. They are accepting that I am available during work hours, and that's it. People will get their answers, and they will get their support, but only when I am good and ready.

So, remember, work hard during the hours of the day you are at work and relax hard during the hours of the day you are not. Unwinding is the best thing you can do and not just for you but for the company you work for.

A Final Word

Thanks for taking the time to read my book. It truly means the world that you picked up my work; hopefully, you have learned a few things.

Writing a book is a labor of love, and readers like you bring it to life. Your curiosity and willingness to understand this topic makes all the late nights and challenges worthwhile.

My ultimate goal as an author is to educate anyone in a similar position to me as I did back when my journey began. Knowing that I've made even a small impact on you fills me with gratitude.

Once again, thank you from the bottom of my heart for reading my book. Your support and enthusiasm mean everything to me. Who knows, maybe our paths will cross one day, and we can have an open discussion about the topics covered in this book.

Made in the USA
Monee, IL
04 November 2026

67385561R00134